Charles Lyell and Moder

by Thomas George B

PREFACE.

The life of Charles Lyell is singularly free from "moving accidents by flood and field." Though he travelled much, he never, so far as can be ascertained, was in danger of life or limb, of brigand or beast. At home his career was not hampered by serious difficulties or blocked by formidable obstacles; not a few circumstances were distinctly favourable to success. Thus his biography cannot offer the reader either the excitement of adventure, or the interest of an unwearied struggle with adverse conditions. But for all that, as it seems to me, it can teach a lesson of no little value. Lyell, while still a young man, determined that he would endeavour to put geology--then only beginning to rank as a science--on a more sound and philosophical basis. To accomplish this purpose, he spared no labour, grudged no expenditure, shrank from no fatigue. For years he was training himself by observation and travel; he was studiously aiming at precision of thought and expression, till "The Principles of Geology" had been completed and published. But even then, though he might have counted his work done, he spared no pains to make it better, and went on at the task of improvement till the close of his long life.

My chief aim, in writing this little volume, has been to bring out this lesson as strongly and as clearly as possible. I have striven to show how Charles Lyell studied, how he worked, how he accumulated observations, how each journey had its definite purposes. Accordingly, I have often given his words in preference to any phrases of my own, and have quoted freely from his letters, diaries, and books, because I wished to show exactly how things presented themselves to his eyes, and how ideas were maturing in his mind. Regarded in this light, Lyell's life becomes an apologue, setting forth the beneficial results of concentrating the whole energy on one definite object, and the moral grandeur of a calm, judicial, truth-seeking spirit.

In writing the following pages I have, of course, mainly drawn upon the "Life, Letters, and Journals," edited by Mrs. Lyell; but I have also made use of his books, especially the "Principles of Geology," and the two tours in North America. I am under occasional obligations to the excellent life, contributed by Professor G. A. J. Cole to the "Dictionary of National Biography," and have to thank my friend Professor J. W. Judd for some important details which he had learnt through his intimacy with the veteran geologist. He also kindly lent the engraving (executed in America from a daguerreotype) which has been copied for the frontispiece of this volume.

T. G. BONNEY.

CONTENTS.

CHAPTER

I.--CHILDHOOD AND SCHOOLDAYS

II.--UNDERGRADUATE DAYS 19

III.--THE GROWTH OF A PURPOSE

IV.--THE PURPOSE DEVELOPED AND ACCOMPLISHED

V.--THE HISTORY AND PLACE IN SCIENCE OF THE "PRINCIPLES OF GEOLOGY"

VI.--EIGHT YEARS OF QUIET PROGRESS

VII.--GEOLOGICAL WORK IN NORTH AMERICA

VIII.--ANOTHER EPOCH OF WORK AND TRAVEL

IX.--STEADY PROGRESS

X.--THE ANTIQUITY OF MAN

XI.--THE EVENING OF LIFE

XII.--SUMMARY

CHARLES LYELL AND MODERN GEOLOGY.

CHAPTER I.

CHILDHOOD AND SCHOOLDAYS.

Caledonia, stern and wild, may be called "meet nurse" of geologists as well as of poets. Among the most remarkable of the former is Charles Lyell, who was born in Forfarshire on November 14th, 1797, at Kinnordy, the family mansion. His father, who also bore the name of Charles,[1] was both a lover of natural history and a man of high culture. He took an interest at one time in entomology, but abandoned this for botany, devoting himself more especially to the study of the cryptogams. Of these he discovered several new species, besides some other plants previously unknown in the British flora, and he contributed the article on Lichens to Smith's "English Botany." More than one species was named after him, as well as a genus of mosses, Lyellia, which is chiefly found in the Himalayas. Later in his life, science, on the whole, was supplanted by literature, and he became engrossed in the study of the works of Dante, of some of whose poems[2] he published translations and notes. Thus the geologist and author is an instance of "hereditary genius."

Charles was the eldest of a family of ten--three sons and seven daughters, all of whom grew up. Their mother was English, the daughter of Thomas Smith, of Maker Hall in Yorkshire, "a woman of strong sense and tender anxiety for her children's welfare." "The front of heaven," as Lyell has written in a fragment of autobiography, was not "full of fiery shapes at his nativity," but the season was so exceptionally warm that his mother's bedroom-window was kept open all the night--an appropriate birth-omen for the geologist, who had a firmer faith than some of his successors in the value of work in the open air. He has put on record only two characteristics of his infancy, and as these can hardly be personal recollections, we may assume them to have been sufficiently marked to impress others. One if not both was wholly physical. He was very late in cutting his teeth, not a single one having appeared in the first twelvemonth, and the hardness of his infant gums caused an old wife to prognosticate that he would be edentulous. Also, his lungs were so vigorous and so habitually exercised that he was pronounced "the loudest and most indefatigable squaller of all the brats of Angus."

The geologist who so emphatically affirmed the necessity of travel, early became an unconscious practiser of his own precept. When he was three months old his parents went from Kinnordy to Inveraray, whence they journeyed to the south of England, as far as Ilfracombe. From this place they removed to Weymouth and thence to Southampton. More than a year must have been thus spent, for their second child--also a son--was born at the last-named town. Mr. Lyell, the father, now took a lease of Bartley Lodge, on the New Forest--some half-dozen miles west of Southampton, where the family lived for twenty-eight years. His mother and sisters also left Kinnordy, and rented a house in Southampton. Their frequent excursions to Bartley Lodge, as Lyell observes, were always welcome to the children, for they never came empty-handed.

Kinnordy, however, was visited from time to time in the summer, and on one of these occasions, when Charles was in his fifth year, some of the family had a narrow escape. They were about a stage and a half from Edinburgh; the parents and the two boys in one carriage; two nursemaids, the cook, and the two youngest children, sisters, in a chaise behind. The horses of this took fright on a narrow part of the road and upset the carriage over a very steep slope. Fortunately all escaped unhurt, except one of the maids, whose arm was cut by the splintered glass. The parents ran to the rescue. "Meanwhile, Tom and I were left in the carriage. We thought it fine pastime, and I am accused of having prompted Tom to assist in plundering the pockets of the carriage of all the buns and other eatables, which we demolished with great speed for fear of interruption."[3] This adventure, however, was not quite his earliest reminiscence; for that was learning the alphabet when he was about three years old.

Charles was kept at home till he had nearly completed his eighth year, when he was sent with his brother Tom to a boarding-school at Ringwood. The master was the Rev. R. S. Davies; the lads were some fifty in number, the Lyells being about the youngest. They seem, however, not to have been ill-treated, though their companions were rather a rough lot, and they were petted by the schoolmaster's daughter. The most sensational incident of his stay at Ringwood was a miniature "town and gown" row, a set fight between the lads of the place and of the school, from which, however, the Lyells were excluded as too young to share in the joys and the perils of war. But the fray was brought to a rather premature conclusion by the joint intervention of foreign powers--the masters of the school and the tradesmen of the town. In those days smuggling was rife on the south coast, and acting the part of revenue officers and contrabandists was a favourite school game; doubtless the more popular because it afforded a legitimate pretext for something like a fight. The fear of a French invasion also kept this part of England on the qui vive, and Lyell well remembered the excitement caused by a false alarm that the enemy had landed. He further recollected the mingled joy and sorrow which were caused by the victory of Trafalgar and the death of Nelson.

The brothers remained at Ringwood only for about two years, for neither the society nor the instruction could be called first-class; and they were sent, after a rather long holiday at home, to another school of about the same size, but much higher character, in Salisbury. The master, Dr. Radcliffe, an Oxford man, was a good classical scholar, and his pupils came from the best families in that part of England. In one respect, the young Lyells found it a change for the worse. At Ringwood they had an ample playground, close to which was the Avon, gliding clear and cool to the sea, a delightful place for a bathe. In a few minutes' walk from the town they were among pleasant lanes; in a short time they could reach the border of the New Forest. But at Salisbury the school was in the heart of the town, its playground a small yard surrounded by walls, and, as he says, "we only walked out twice or three times in a week, when it did not rain, and were obliged to keep in ranks along the endless streets and dusty roads of the suburbs of a city. It seemed a kind of prison by comparison, especially to me, accustomed to liberty in such a wild place as the New Forest." One can sympathise with his feelings, for a procession of schoolboys, walking two and two along the streets of a town, is a dreary spectacle.

But an occasional holiday brought some comfort, for then they were sent on a longer excursion. The favourite one was to the curious earthworks of Old Sarum, then in its glory as a "rotten borough," one alehouse, with its tea-gardens attached, sending two members to Parliament. On these excursions more liberty seems to have been permitted. The boys broke up the large flints that lay all about the ground, to find in them cavities lined with chalcedony or drusy crystals of quartz. But the chief interest centred around a mysterious excavation in the earthwork, "a deep, long subterranean tunnel, said to have been used by the garrison to get water from a river in the plain below." To this all new-comers were taken to listen to the tale of its enormous depth and subterranean pool. Then, when duly overawed, they felt their hats fly off their heads and saw them rolling out of sight down the tunnel. An interval followed of blank dismay, embittered, no doubt, by dismal anticipations of what would probably happen when they got back to the school-house. Then one of the older boys volunteered to act the sybil and lead the way to the nether world. Of course they "regained their felt and felt what they regained"--literally, for the hole was dark enough, though we may set down the "many hundred yards" (which Lyell says that he descended before he recovered his lost hat) as an instance of the permanent effect of a boyish illusion on even a scientific mind.

But the restrictions of Salisbury made the liberty of the New Forest yet more dear. Bartley was an ideal home for boys. It was surrounded by meadows and park-like timber. A two-mile walk brought the lads to Rufus Stone, and on the wilder parts of the Forest. There they could ramble over undulating moors, covered with heath and fern, diversified by marshy tracts, sweet with bog-myrtle, or by patches of furze, golden in season with flowers; or they could wander beneath the shadows of its great woods of oak and beech, over the rustling leaves, among the flickering lights and shadows, winding here and there among tufts of holly scrub, always led on by the hope of some novelty-- a rare insect fluttering by, a lizard or a snake gliding into the fern, strange birds circling in the air, a pheasant or even a woodcock springing up almost under the feet. The rabbits scampered to their holes among the furze; a fox now and again stole silently away to cover, or a stag--for the deer had not yet been destroyed--was espied among the tall brake. Those, too, it must be remembered, were the days when boys got their holidays in the prime of the summer, at the season of haymaking and of ripe strawberries. They were not kept stewing in hot school-rooms all through July, until the flowers are nearly over and the bright green of the foliage is dulled, until the romance of the summer's youth has given place to the dulness of its middle age. In these days it is our pleasure to do the right thing in the wrong place--a truly national characteristic. We all--young and old--toil through the heat and the long days, and take holiday when the autumn is drawing nigh and Nature writes "Ichabod" on the beauty of the waning year.

At Salisbury, Lyell had two new experiences--the sorrows of the Latin Grammar and the joys of a bolster-fight. But his health was not good; a severe attack of measles in the first year was followed in the second by a general "breakdown," with symptoms of weakness of the lungs. So he was taken home for three months to recruit. This was at first a welcome change from the restrictions of Salisbury; but, as his lessons necessarily were light, he began to mope for want of occupation; for, as he says, "I was always most exceedingly miserable if unemployed, though I had an excessive aversion to work unless forced to it." So he began to collect insects--a pursuit which, as he remarks, exactly suited him, for it was rather desultory, gave employment to both mind and body, and gratified the "collecting" instinct, which is strong in most boys. He began with the lepidoptera, but before long took an interest in other insects, especially the aquatic. Fortunately his father had been for a time a collector, and possessed some good books on entomology, from the pictures in which Charles named his captures. This was, of course, an unscientific method, but it taught him to recognise the species and to know their habits. There are few better localities for lepidoptera, as every collector knows, than the New Forest, and some of the schoolboy's "finds" afterwards proved welcome to so well known an entomologist as Curtis. But when Charles returned to school he had to lay aside, for a season, the new hobby; for in those days a schoolboy's interest in natural history did not extend beyond birds'-nesting, and his little world was not less, perhaps even more frank and demonstrative than now, in its criticism of any innovation or peculiarity on the part of one of its members.

The school at Salisbury appears to have been a preparatory one, so before very long another had to be sought. Mr. Lyell wished to send his two boys to Winchester, but found to his disappointment that there would not be a vacancy for a couple of years; so after instructing them at home for six months, he contented himself with the Grammar School at Midhurst, in Sussex, at the head of which was one Dr. Bayley, formerly an under-master at Winchester. Charles, now in his thirteenth year, found this, at first, a great change. The school contained about seventy boys, big as well as little, and its general system resembled that of one of the great public schools. He remarks of this period of his life: "Whatever some may say or sing of the happy recollections of their schooldays, I believe the generality, if they told the truth, would not like to have them over again, or would consider them as less happy than those which follow." He was not the kind of boy to find the life of a public school very congenial. Evidently he was a quietly-disposed lad, caring more for a country ramble than for games; perhaps a little old-fashioned in his ways; not pugnacious, but preferring a quiet life to the trouble of self-assertion. So, in his second half-year, when he was left to shift entirely for himself, his life was "not a happy one," for a good deal of the primeval savage lingers in the boys of a civilised race. It required, as he said, a good deal to work him up to the point of defending his independence; thus he was deemed incapable of resistance and was plagued accordingly. But at last he turned upon a tormentor, and a fight was the result. It was of Homeric proportions, for it lasted two days, during five or six hours on each, the combatants being pretty evenly matched; for though Lyell's adversary was rather the smaller and weaker, he knew better how to use his fists. Strength at the end prevailed over science, though both parties were about equally damaged. The vanquished pugilist was put to bed, being sorely bruised in the visible parts. Lyell, whose hurts were mostly hidden, made light of them, by the advice of friends, but he owns that he ached in every bone for a week, and was black and blue all over his body. Still he had not fought in vain, for, though the combat won him little honour, it delivered him from sundry tormentors.

The educational system of the school stimulated his ambition to rise in the classes. "By this feeling," he says, "much of my natural antipathy to work, and extreme absence of mind, was conquered in a great measure, and I acquired habits of attention which, however, were very painful to me, and only sustained when I had an object in view." There was an annual speech-day, and Charles, on the first occasion, obtained a prize for his performance. "Every year afterwards," he continues, "I received invariably a prize for speaking, until high enough to carry off the prizes for Latin and English original composition. My inventive talents were not quick, but to have any is so rare a qualification that it is sure to obtain a boy at our great schools (and afterwards as an author) some distinction." Evidently he gave proofs of originality beyond his fellows; since he won a prize for English verse, though he had written in the metre of the "Lady of the Lake" instead of the ordinary ten-syllabic rhyme. On another occasion he commemorated, in his weekly Latin copy, the destruction of the rats in a neighbouring pond, writing in mock heroics, after the style of Homer's battle of the frogs and mice.

The school, like all other collections of boys, had its epidemic hobbies. The game of draughts, coupled unfortunately with gambling on a small scale, was followed by chess, and that by music. To each of these Charles was more or less a victim, and his progress up the school was not thereby accelerated.

Birds'-nesting also had a turn in its season. His love for natural history made him so keen in this pursuit that he became an expert climber of trees. But his schooldays on the whole were uneventful, and he went to Oxford at a rather early age, his brother Tom having already left Midhurst in order to enter the Navy.

FOOTNOTES:

[1] Born 1767, died 1849 (also son of a Charles Lyell); educated at St. Andrew's and at St. Peter's College, Cambridge, where he proceeded to the degree of B.A. in 1791 and M.A. in 1794.

[2] In 1835, the Canzoniere, including the Vita Nuova and Convito; a second edition was published in 1842; in 1845 a translation of the Lyrical Poems of Dante.

[3] Life, Letters, and Journals, vol. i. p. 3.

CHAPTER II.

UNDERGRADUATE DAYS.

Lyell matriculated at Exeter College, and appears to have begun residence in January, 1816--that is, soon after completing his eighteenth year. At Oxford, though not a "hard reader," he was evidently far from idle, and wrote for some of the University prizes, though without success. Several of his letters to his father have been preserved. In these he talks about his studies, mathematical and classical; criticises Coleridge's "Christabel," and praises Kirke White's poetry; describes the fritillaries blossoming in the Christchurch meadows, and refers occasionally to political matters. The letters are well expressed, and indicate a thoughtful and observant mind. While yet a schoolboy he had stumbled upon a copy of Bakewell's "Geology" in his father's library, which had so far awakened his interest that in the earlier part of his residence at Oxford he attended a course of Professor Buckland's lectures, and took careful notes. The new study is briefly mentioned in a letter, dated July 20th, 1817. This is written from Yarmouth, where he is visiting Mr. Dawson Turner, the well-known antiquarian and botanist. He states that, on his way through London, he went to see the elephant at Exeter

Change, Bullock's Museum, and Francillon's collection of insects. At Norwich also he saw more insects, the cathedral, and some chalk pits, in which he found an "immense number of belemnites, echinites, and bivalves." He was also greatly interested by the fossils in Dr. Arnold's collection at Yarmouth, particularly by the "alcyonia" found in flints.[4] A few days later he again dwells on geology, and speculates shrewdly on the formation of the lowland around Yarmouth and the ancient course of the river. In one paragraph a germ of the future "Principles" may be detected. It runs thus:

"Dr. Arnold and I examined yesterday the pit which is dug out for the foundation of the Nelson monument, and found that the first bed of shingle is eight feet down. Now this was the last stratum brought by the sea; all since was driven up by wind and kept there by the 'Rest-harrow' and other plants. It is mere sand. Therefore, thirty-five years ago the Deens were nearly as low as the last stratum left by the sea; and as the wind would naturally have begun adding from the very first, it is clear that within fifty years the sea flowed over that part. This, even Mr. T. allows, is a strong argument in favour of the recency of the changes. Dr. Arnold surprised me by telling me that he thought that the Straits of Dover were formerly joined, and that the great current and tides of the North Sea being held back, the sea flowed higher over these parts than now. If he had thought a little more he would have found no necessity for all this, for all those towns on this eastern coast, which have no river god to stand their friend, have necessarily been losing in the same proportion as Yarmouth gains--viz. Cromer, Pakefield, Dunwich, Aldborough, etc., etc. With Dunwich I believe it is Fuit Ilium."[5]

Evidently Lyell by this time had become deeply interested in geology, for his journal contains several notes made on the road from London to Kinnordy, and records, during his stay there, not only the capture of insects, but also visits to quarries, and the discovery of crystallised sulphate of barytes at Kirriemuir and elsewhere.

Towards the end of his first long vacation he travelled, in company with two friends of his own age, from Forfarshire across by Loch Tay, Tyndrum, and Loch Awe, to the western coast at Oban, whence they visited Staffa and Iona. With the caves in the former island he was greatly impressed; and he noted the columns of basalt, which, he said, were "pentagonal" in form, quite different from the "four-square" jointing of the red granite at the south-west end of Mull. With the ruins of Iona he was a little disappointed, for he wrote in his diary that "they are but poor after all." The wonders of Fingal's Cave appealed to his poetical as well as to his geological instincts, for in October, after his return to Oxford, he sent to his father some stanzas on this subject which are not without a certain merit. But the covering letter was mostly devoted to geology.

The next year, 1818, marked an important step in his education as a geologist, for he accompanied his father, mother, and two eldest sisters on a Continental tour. Starting early in June, they drove in a ramshackle carriage, which frequently broke down, from Calais to Paris, along much the same route as the railway now takes; they visited the sights of the capital, not forgetting either the artistic treasures of the Louvre or the collections of the Jardin des Plantes, particularly the fossils of the "Paris basin." Thence they journeyed by Fontainebleau and Auxerre and he makes careful and shrewd notes on the geology, for the carriage travelling of those days, though slow, was not without its advantages--and in crossing the Jura he observes the nodular flints in a limestone, and the contrast between these mountains and the Grampians of his native land. As they descended the well-known road which leads down to Gex in Switzerland, they had the good fortune to obtain a splendid view of Mont Blanc and the Alps. From Geneva, where he notes the "most peculiar deep blue colour of the Rhone," they visited Chamouni by the usual route. At this time the principal glaciers were advancing rather rapidly. The Glacier des Bossons, he remarks, "has trodden down the tallest pines with as much ease as an elephant could the herbage of a meadow. Some trunks are still seen projecting from the rock of ice, all the heads being embodied in this mass, which shoots out at the top into tall pyramids and pinnacles of ice, of beautiful shapes and a very pure white.... It has been pressed on not only through the forest, but over some cultivated fields, which are utterly lost."[6]

At Chamouni, Lyell made the most of his time, for in three days he walked up to the Col de Balme, climbed the Brent, and made his first glacier expedition, to the well-known oasis among the great fields of snow and ice which is called the Jardin. Everywhere he notes the flowers, which at that season were in full beauty; and the insects, capturing "no less than seven specimens of that rare insect, Papilio Apollo."[7] He feels all the surprise and all the delight which thrills the entomologist from the British Isles when he first sets foot on the slopes of the higher Alps, and sees in abundance the rarities of his own country, besides not a few new species. But Lyell does not neglect the rocks and minerals, or the red snow, or the wonders of the ice world. Chamouni, we are told, was then "perfectly inundated with English," for fifty arrived in one day. The previous year they had numbered one thousand out of a total of fourteen hundred visitors. Since then, times and the village have changed.

Returning to Geneva, the party travelled by Lausane and then followed the picturesque route along the river, by the tumultuous rapids of Laufenburg and the grand falls of the Rhine, to Schaffhausen, whence they turned off to Zurich. Here he writes of the principal inn that it "partook more than any of a fault too common in Switzerland. They have their stables and cow-houses under the same roof, and the unavoidable consequences may be conceived, till they can fall in with a man as able--as 'Hercules to cleanse a stable.'"

From Zurich they crossed the Albis to Zug. The other members of the party went direct to Lucerne, but Lyell turned aside to visit the spot where twelve years previously an enormous mass of pudding-stone had come crashing down from the Rossberg, had destroyed the village of Goldau, and had converted a great tract of fertile land into a wilderness of broken rock. He diagnosed correctly the cause of the catastrophe, and then ascended the Rigi. Here he spent a flea-bitten night at the Kulm Hotel, but was rewarded by a fine sunset and a yet finer sunrise.

At Lucerne he rejoined his relatives, and they drove together to Meyringen. From this place they made an excursion to the Giessbach Falls, and saw the Alpbach in flood after a downpour of rain. This, like some other Alpine streams, becomes at such times a raging mass of liquid mud and shattered slate, and Lyell carefully notes the action of the torrent under these novel circumstances, and its increased power of transport. Parting from his relatives at the Handeck Falls, he walked up the valley of the Aar to the Grimsel Hospice, where he passed the night, and the next morning crossed over into the valley of the Rhone to the foot of its glacier, and then walked back again to Meyringen. He remarks that on the way to the Hospice "we passed some extraordinary large bare planks of granite rock above our track, the appearance of which I could not account for." This is not surprising, for he had not yet learnt to read the "handwriting on the wall" of a vanished glacier.

Its interpretation was not to come for another twenty years, when these would be recognised as perhaps the finest examples of ice-worn rocks in Switzerland. Lyell was evidently a good pedestrian; for the very next day he walked from Meyringen over the two Scheideggs to Lauterbrunnen, ultimately joining his relatives at Thun, from which town they went on to Berne, where they were so fortunate as to see, from the well-known terrace, the snowy peaks of the Oberland in all the beauty of the sunset glow.

Then they journeyed over the pleasant uplands to Vevay, and so by the shore of the Lake of Geneva and the plain of the Rhone valley to Martigny, turning aside to visit the salt mines near Bex. They reached Martigny a little more than seven weeks after the lake, formed in the valley of the Dranse by the forward movement of the Glacier, had burst its icy barrier, and they saw everywhere the ruins left by the rush of the flood. The road as they approached Martigny was even then, in some places, under water; in others it was completely buried beneath sand. The lower storey of the hotel had been filled with mud and debris, which was still piled up to the courtyard. Lyell went up the valley of the Dranse to the scene of the catastrophe, and wrote in his journal an interesting description of both the effects of the flood and the remnants of the ice-barrier. Before returning to Martigny he also walked up to the Hospice on the Great St. Bernard, and then the whole party crossed by the Simplon Pass into Italy, following the accustomed route and visiting the usual sights till they arrived at Milan.

The next stage on their tour--and this must have been in those days a little tedious--brought them to Venice. The Campanile Lyell does not greatly admire, and of St. Mark's he says rather oddly, "The form is very cheerful and gay"; but on the whole he is much impressed with the buildings of Venice, and especially with the pictures. On their return they went to Bologna, and then crossed the Apennines to Florence. Everywhere little touches in the diary indicate a mind exceptionally observant--such as notes on the first firefly, the fields of millet, the festooned vines seen on the plain, or the peculiar sandy zone on the northern slopes of the hills. He also mentions that shortly after crossing the frontier of Tuscany they passed near Coviliajo, "a volcanic fire" which proceeded from a neighbouring mountain.[8] This they intended to visit on their return. But at Florence the diary ends abruptly, for the note-book which contained the rest of it was unfortunately lost.

We have given this summary of Lyell's journal in some detail, but even thus it barely suffices to convey an adequate idea of the cultured tastes, wide interests, and habits of close and accurate observation disclosed by its pages. It shows, better perhaps than any other documents, the mental development of the future author of the "Principles of Geology." Few things, as he journeys, escape his notice; he describes facts carefully and speculates but little. As he wanders among the Alpine peaks, he makes no reference to convulsions of the earth's crust; as he views the ruin wrought by the Dranse, he says naught of deluges.

The travellers got back to England in September, and at the end of the Long Vacation Lyell returned to Oxford. There he remained till December, 1819, when he proceeded to the degree of Bachelor of Arts, obtaining a second class in Classical Honours. Considering that he had never been a "hard reader," and that he appears to have spent much of his "longs" in travel--a practice which, though good for general education, counts for little in the schools--the position indicates that he possessed rather exceptional abilities and a good amount of scholarship. Though Oxford had been unable to bestow upon him a systematic training in science, she had given a definite bias to his inclination, and had fostered and cultivated a taste for literature which in the future brought forth a rich fruitage.

FOOTNOTES:

[4] Probably they were fossil sponges.

[5] Life, Letters, and Journals, vol. i. p. 43.

[6] Life, Letters, and Journals, vol. i. p. 69.

[7] Now generally called Parnassius Apollo; but very likely he captured more than one species of the genus.

[8] Probably it was a bituminous shale which had become ignited, as was the case at Ringstead Bay, Dorset, with the Kimeridge clay. The same often happens with the "banks" of coal-pits.

CHAPTER III.

THE GROWTH OF A PURPOSE.

Shortly after he had donned the bachelor's hood Lyell came to London, was entered at Lincoln's Inn, and studied law in the office of a special pleader. Science was not forsaken, for in March, 1819, he was elected a Fellow of the Geological Society, and about the same time joined the Linnean Society. Before very long his legal studies were interrupted. His eyes became so weak that a complete rest was prescribed; accordingly, in the autumn of 1820, he accompanied his father on a journey to Rome. During this but little was done in geology, for the travellers spent almost all their time in towns.

On his return, so far as can be inferred from the few letters which have been published, Lyell continued to work at geology, and at Christmas, 1821, was seeking in vain for freshwater fossils in the neighbourhood of Bartley. In the spring of 1822 he investigated the Sussex coast from Hastings to Dungeness, and studied the effects of the sea at Winchelsea and Rye. In the early summer of 1823 he visited the Isle of Wight, and in a letter to Dr. Mantell suggested that the "blue marl"[9] in Compton Chine is identical with that at Folkestone, and compared the underlying strata with those in Sussex, clearing up some confusions, into which earlier observers had fallen, about the Wealden and Lower Greensand. He was now evidently beginning to get a firm grip on the subject--a thing far from easy in days when so little had been ascertained-- and this year he read his first papers to the Geological Society--one, in January, written in conjunction with Dr. Mantell, "On the Limestone and Clay of the Ironsand in Sussex"; the other in June, "On the Sections presented by Some Forfarshire Rivers." Also, on February 7th, he was elected one of the secretaries of that Society, an office which he retained till 1826. This is a pretty clear proof that he had begun to make his mark among geologists, and was well esteemed by the leaders of the science.

No sooner had he returned from the Isle of Wight than he started for Paris, going direct from London to Calais, in the Earl of Liverpool steam packet, "in 11 hours! 120 miles! engines 80 horse-power for 240 tons." In the last letter written to his father before quitting England he refers to our neighbours across the Channel in the following terms: "My opinion of the French people is that they are much too corrupt for a free government and much too enlightened for a despotic one." That was written full seventy years ago;

perhaps even now, were he alive, he would not be disposed to withdraw the words.

At Paris he was well received by Cuvier, Humboldt, and other men of science, attended lectures at the Jardin du Roi, and saw a good deal of society. His letters home often contain interesting references to matters political and social--such as, for example, the following remarks which he heard from the mouth of Humboldt: "You cannot conceive how striking and ludicrous a feature it is in Parisian society at present that every other man one meets is either minister or ex-minister. So frequent have been the changes. The instant a new ministry is formed, a body of sappers and miners is organised. They work industriously night and day. At last the ministers find that they are supplanted by the very arts by which a few months ago they raised themselves to power."[10] Lyell more than once expresses a regret, which, indeed, was generally felt in scientific circles, that Cuvier had lost caste by "dabbling so much with the dirty pool of politics"; and himself works away at geology, studying the fossils of the Paris basin in the museums, and visiting the most noted sections in order to add to his own collection and observe the relations of the strata.

He returned to England towards the end of September, and no doubt spent the next few months in working at geology as far as his eyes, which were becoming stronger, permitted. The summer of 1824 was devoted to geological expeditions. In the earlier part he took Mons. Constant Prost, one of the leaders of geology in France, to the west of England. Their special purpose was to examine the Jurassic rocks, but they extended their tour as far as Cornwall. Afterwards Lyell went to Scotland, where he was joined by Professor Buckland; and the two friends, after spending a few days in Ross-shire, went to Brora, and then returned from Inverness by the Caledonian canal. This gave them the opportunity of examining the famous "parallel roads" of Glenroy, which were the more interesting because they had already seen something of the kind near Cowl, in Ross-shire. Afterwards they went up Glen Spean and crossed the mountains to Blair Athol, visiting the noted locality in Glen Tilt, where Hutton made his famous discovery of veins of granite intrusive in the schists of that valley, and then they made their way to Edinburgh. Here much work was done, both among collections and in the field, and it was lightened--as might be expected in a place so hospitable--by social pleasures and friendly converse with some of the leading literary and scientific men.

Four years of comparative rest and frequent change of scene had produced such an improvement in the condition of his eyes that he was able to resume his study of the law, and was called to the Bar in 1825. For two years he went on the Western Circuit, having chambers in the Temple and getting a little business. But, as his correspondence shows, geology still held the first place in his affections,[11] and papers were read to the Society from time to time. Among them one of the most important, though it was not printed in their journal, described a dyke of serpentine which cut through the Old Red Sandstone on the Kinnordy estate.[12] But, as is shown by a letter to his sister, written in the month of November, he had not lost his interest in entomology. At that time the collectors of insects in Scotland were very few in number, and the English lepidopterists welcomed the specimens which Lyell and his sister had caught in Forfarshire. The family had left Bartley Lodge in the earlier part of the year and had settled in the old home at Kinnordy. About this time also Lyell began to contribute to the Quarterly Review, writing articles on educational and scientific topics. This led to a friendship with Lockhart, who became editor at the end of 1825, and gave him an introduction to Sir Walter Scott. A Christmas visit to Cambridge introduced him to the social life of that university.

In the spring of 1827 his ideas as to his future work appear to have begun to assume a definite form. To Dr. Mantell[13] he writes that he has been reading Lamarck, and is not convinced by that author's theories of the development of species, "which would prove that men may have come from the ourang-outang," though he makes this admission: "After all, what changes a species may really undergo! How impossible will it be to distinguish and lay down a line, beyond which some of the so-called extinct species have never passed into recent ones!" The next sentence is significant: "That the earth is quite as old as he [Lamarck] supposes has long been my creed, and I will try before six months are over to convert the readers of the Quarterly to that heterodox opinion."[14] A few lines further on come some sentences which indicate that the leading idea of the "Principles" was even then floating in his mind. "I am going to write in confirmation of ancient causes having been the same as modern, and to show that those plants and animals, which we know are becoming preserved now, are the same as were formerly." Hence, he proceeds to argue, it is not safe to infer that because the remains of certain classes of plants or animals are not found in particular strata, the creatures themselves did not then exist. "You see the drift of my argument," he continues; "ergo, mammalia existed when the oolite and coal, etc., were formed."[15] The first of these quotations strikes the keynote of modern geology as opposed to the older notions of the science; what follows suggests a caution, to which Darwin afterwards drew more particular attention, though he turned the weapon against Lyell himself, viz. "the imperfection of the geological record."

A letter to his father, also written in the month of April, shows that, while he has an immediate purpose of opening fire on MacCulloch,[16] who had bitterly attacked in the Westminster Review Scrope's book upon Volcanoes, he has "come to the conclusion that something of a more scientific character is wanted, for which the pages of a periodical are not fitted." He might, he says, write an elementary book, like Mrs. Marcet's "Conversations on Chemistry," but something on a much larger scale evidently is floating on his mind. In this letter also he discusses his prospects with his father, who apparently had suggested that he should cease from going on circuit; and argues that he gains time by appearing to be engaged in a profession, for "friends have no mercy on the man who is supposed to have some leisure time, and heap upon him all kinds of unremunerative duties." Lyell was not devoid of Scotch shrewdness, and doubtless early learnt that when it is all work and no pay men see your merits through a magnifying glass, but when it comes to the question of a reward, they shift the instrument to your defects.

Gradually the plan of the future book assumed a more definite shape in his mind, as we can see from a letter to Dr. Mantell early in 1828. About this time also Murchison, with whom he was planning a long visit to Auvergne,[17] appears among his correspondents. Herschel[18] tells him how he and Faraday had melted in a furnace "granite into a slag-like lava"; Hooker[19] begs him to notice the connection between plants and soils as he travels; his father urges him to take his clerk with him to act as amanuensis and save his eyes, which might be affected by the glare of the sun, and to help him generally in collecting specimens and carrying the barometers. Early in the month of May he started for Paris, where he met Mr. and Mrs. Murchison, and the party left for Clermont Ferrand in a "light open carriage, with post horses." As far as Moulins the roads were bad, but as they receded from Paris and approached the mountains "the roads and the rates of posting improved, so that we averaged nine miles an hour, and the change of horses [was] almost as quick as in England. The politeness of the people has much delighted us, and they are so intelligent that we get much geology from them." Clermont Ferrand became their headquarters for some time, and Lyell's letters to his father are full of notes on the geology of the district, one of the most interesting in Europe. The great plateau which rises on the western side of the broad valley of the Allier is studded with cones and craters--some so fresh that one might imagine their last eruptions to have happened during the decline of the Roman empire;[20] others in almost every stage of dissection by the scalpels of nature. Streams of lava, still rough and clinkery, have poured themselves over the plateau and have run down the valleys till they have reached the plain of the Allier, while huge fragments of flows far larger and more ancient have been carved by the action of rain and rivers into natural bastions, and now may be seen resting upon stratified marls, crowded with freshwater shells and other organisms,--the remnants of deposits accumulated in great lakes, which had been already drained in ages long before man appeared on the earth.

The two geologists worked hard, for who could be idle in such a country as this? They often began at six in the morning and rested not till evening, though the summers are hot in Auvergne, and this one was exceptionally so. Lyell writes home, "I never did so much real geology in so many days." Mrs. Murchison also was "very diligent, sketching, labelling specimens, and making out shells, in which last she is a valuable assistant." Sometimes they went farther afield, visiting Pontgibaud and the gorge of the Sioul, where they found a section previously unnoticed, which gave them a clear proof that a lava-stream had dammed up the course of a river by flowing down into its valley, and had converted the part above into a lake. This again had been drained as the river had carved for itself a new channel, partly in the basalt, partly in the underlying gneiss. Here, then, was a clear proof that a river could cut out a path for itself, and that forces still in operation were sufficient, given time enough, to sculpture the features of the earth's crust. Notwithstanding the hard work, the outdoor life suited Lyell, who writes that his "eyes were never in such condition before." Murchison, too, was generally in good health, but would have been better, according to his companion, if he had been a little more abstemious at table and a worse customer to the druggist.

From Clermont Ferrand the travellers moved on to the Cantal, where they investigated the lacustrine deposits beneath the lava-streams all around Aurillac. These deposits exhibited on a grand scale the phenomena which Lyell had already observed on a small one in the marls of the loch at Kinnordy. Thence they went on through the Ard阢he and examined the "pet volcanoes of the Vivarais," as they had been termed by Scrope. The Murchisons now began to suffer from the heat, for it was the middle of July. Nevertheless, they still pushed on southwards, and after visiting the old towns of Gard and the Bouches du Rhue, went along the Riviera to Nice, having been delayed for a time at Frus, where Murchison had a sharp attack of malarious fever. It was an exceptionally dry summer, and the town in consequence was malodorous; so after a short halt, they moved on to Milan and at last arrived at Padua, working at geology as they went along, and constantly accumulating new facts. From Padua they visited Monte Bolca, noted for its fossil fish, the Vicentin, with its sheets of basalt, and the Euganean Hills, where the "volcanic phenomena [were] just Auvergne over again." Then the travellers parted, the Murchisons turning northward to the Tyrol, while Lyell continued on his journey southward to Naples and Sicily.

Some four months had now been spent, almost without interruption, in hard work and the daily questioning of Nature. The results had surpassed even Lyell's anticipations; they had thrown light upon the geological phenomena of the remote past, and cleared up many difficulties which, hitherto, had impeded the path of the investigators. On the coast of the Maritime Alps Lyell had found huge beds of conglomerate, parted one from another by laminated shales full of fossils, most of which were identical with creatures still living in the Mediterranean. These masses attained a thickness of 800 feet, and were displayed in the sides of a valley fifteen miles in length. They supplied a case parallel with that of the conglomerates and sandstones of Angus, and indicated that no extraordinary conditions--no deluges or earth shatterings--had been needed in order to form them. If the torrents from the Maritime Alps, as they plunged into the Mediterranean, could build up these masses of stratified pebbles, why not appeal to the same agency in Scotland, though the mountains from which they flowed, and the sheet of water into which they plunged, have alike vanished? The great flows of basalt--some fresh and intact, some only giant fragments of yet vaster masses--the broken cones of scoria, and the rounded hills of trachyte in Auvergne, had supplied him with links between existing volcanoes and the huge masses of trap with which Scotland had made him familiar; while these basalt flows--modern in a geological sense, but carved and furrowed by the streams which still were flowing in their gorges--showed that rain and rivers were most potent, if not exclusive, agents in the excavation of valleys. "The whole tour," thus he wrote to his father, "has been rich, as I had anticipated (and in a manner which Murchison had not), in those analogies between existing nature and the effects of causes in remote eras which it will be the great object of my work to point out. I scarcely despair now, so much do these evidences of modern action increase upon us as we go south (towards the more recent volcanic seat of action) of proving the positive identity of the causes now operating with those of former times."[21]

One important result of this journey was a conjoint paper on the excavation of valleys in Auvergne, which was written before the friends parted, and was read at the Geological Society in the later part of the year. Lyell writes thus to one of his sisters from Rome, on his return thither, in the following January[22]:--

"My letters from geological friends are very satisfactory as to the unusual interest excited in the Geological Society by our paper on the excavation of valleys in Auvergne. Seventy persons present the second evening, and a warm debate. Buckland and Greenough furious, contra Scrope, Sedgwick, and Warburton supporting us. These were the first two nights in our new magnificent apartments at Somerset House." He adds, "Longman has paid down 500 guineas to Mr. Ure, of Dublin, for a popular work on geology, just coming out. It is to prove the Hebrew cosmogony, and that we ought all to be burnt in Smithfield."

On the way to Naples, Lyell made several halts: at Parma, Bologna, Florence, Siena, Viterbo, and Rome; visiting local geologists, studying their collections of fossil shells, keeping his eye more especially on the relations which the species exhibited with the fauna still existing in the Mediterranean, and losing no opportunity of examining the ancient volcanic vents and the crater lakes, which form in places such remarkable features in the landscape. "The shells in the travertine," he writes, "are all real species living in Italy, so you perceive that the volcanoes had thrown out their ash, pumice, etc., and these had become covered with lakes, and then the valleys had been hollowed out, all before Rome was built, 2,500 years and more ago."

On reaching Naples, he climbed Vesuvius, and saw for the first time the lava-streams and piles of scoria of a volcano still active; while the wonderful sections of the old crater of Somma furnished a link between the living present and the remote past--between Italy and Auvergne. He visited Ischia, where another delightful surprise awaited him, for on its old volcano, Monte Epomeo, he found, at a height of 2,000 feet above the sea, marine shells which belonged "to the same class as those in the lower regions of Ischia." They were contained in a mass of clay, and were quite unaltered. This was a great discovery, for the existence of these fossils "had not been dreamt of," and it showed that the land had been elevated to this extent without any appreciable change in the fauna inhabiting the Mediterranean. Except for this, the island was "an admirable illustration of Mont Dore." He made an excursion also to the Temples, wonderful from the weird beauty of their ruins, on the flat plain between the Apennines and the sea, but with interest geological as well as archeological, because of the blocks of rough travertine with which their columns are built. These he studied, and he visited the quarries from which they were hewn. His letters frequently contain interesting references to the tyranny of the Government, "the inquisitorial suppression of all cultivation of science, whether moral or physical," the idle, happy-go-lucky habits of the common people, the prevalent mendicancy, universal dishonesty, and general corruption. One instance may be worth quoting--it indicates the material with which "United Italy " has had to deal. He wanted to pre-pay the postage of a letter to England. The head waiter at his hotel had said to him, "'Mind, if it is to England you only pay fifteen grains' (sous). I thought the hint a trait of character, as they are all suspicious of one another. The clerk demanded twenty-five. I remonstrated, but he insisted, and, as he was dressed and had the manners of a gentleman, I paid. When I found on my return that I had been cozened, I asked the head waiter, with some indignation, 'Is it possible that the Government officers are all knaves?' 'Sono Napolitani, Signor; la sua eccellenza mi scusera, ma io sono Romano!'"[23] The old proverb, what is bred in the bone will out in the flesh, still holds good; but we may doubt whether the standard of virtue is quite so high as the speaker intimated in certain other provinces which Piedmont has acquired at the price of the cradle of the royal house and some of the best blood of the nation.

At Naples, Lyell was detained longer than he had expected, waiting for a

Government steamer. "There was," he says, "no other way of going, for the pirates of Tripoli have taken so many Neapolitan vessels that no one who has not a fancy to see Africa will venture." But he arrived in Sicily before the end of November, and succeeded in reaching the summit of Etna on the first of December. He was only just in time, for the next day bad weather set in, snow fell heavily, and the summit of the mountain became practically inaccessible for the winter. But as it was, he was able to examine carefully another active volcano, the phenomena of which corresponded with those of Vesuvius, though on a grander scale. From Nicolosi, where he was delayed a day or two by the weather, Lyell went along the Catanian plain to Syracuse and southward to the extreme point of the island, Cape Passaro. From this headland he followed the coast westward as far as Girgenti, and then struck across the island in an easterly direction till he came within about a day's journey of Catania, and then he turned off in a north-westerly direction through the island to Palermo. In this zigzag journey, which occupied about five weeks, he succeeded in obtaining a good general knowledge of the geology of the eastern part of the island; he examined many sections and collected many fossils, thus obtaining material for an accurate classification of the little-known deposits of the Sicilian lowland, and in addition he lost no opportunity of studying the relations of the volcanic masses, wherever they occurred, to the sedimentary strata. As his letters show, bad roads, poor fare, and miserable accommodation made the journey anything but one of pleasure; but its results, as he wrote to Murchison, "exceeded his warmest expectations in the way of modern analogies."

By December 10th he was once more back in the Bay of Naples. As he returned through Rome he availed himself of the opportunity of examining the travertines of Tivoli, which, as he remarked, presented more analogies with those of Sicily than of Auvergne, and welcomed the news that the bones of an elephant had been found in an alluvial deposit which lay beneath the lava of an extinct Tuscan volcano. His notes also prove that he was beginning to see his way to the classification of the extensive deposits of sand and marl in Italy and Sicily, which were subsequently recognised as belonging to the Pliocene era.

Early in February Lyell reached Geneva on his homeward journey, after crossing the Mont Cenis, and by the 19th was back in Paris among his geological friends, "pumping them," as he says, and being well pumped in return. Some of them, he finds, "have come by most opposite routes to the same conclusions as myself, and we have felt mutually confirmed in our views, although the new opinions must bring about an amazing overthrow in the systems which we were carefully taught ten years ago." The accurate knowledge of Deshayes, one of the most eminent conchologists of that day, was especially helpful in bringing his field work in Italy and Sicily into clear and definite order, and he obtained from him a promise of tables of more than 2,000 species of Tertiary shells, from which (he writes to his sister Caroline, who shared his entomological tastes) "I will build up a system on data never before obtained, by comparing the contents of the present with more ancient seas, and the latter with each other."[24]

By the end of February he is back in London and at the Geological Society, defending his views on the constancy of Nature's operations--views which seemed rank heresy to the older school, who sought to solve every difficulty by a convulsion, and were fettered in their interpretation of the records of geology by supposed theological necessities. In April Lyell writes thus to Dr. Mantel[25]:--

"A splendid meeting [at the Geological Society] last night, Sedgwick in the chair. Conybeare's paper on Valley of the Thames, directed against Messrs. Lyell and Murchison's former paper, was read in part. Buckland present to defend the 'Diluvialists,' as Conybeare styles his sect; and us he terms 'Fluvialists.' Greenough assisted us by making an ultra speech on the importance of modern causes.... Murchison and I fought stoutly, and Buckland was very piano. Conybeare's memoir is not strong by any means. He admits three deluges before the Noachian! and Buckland adds God knows how many catastrophes besides; so we have driven them out of the Mosaic record fairly."

Again, in the month of June, he writes to the same correspondent in regard to the second portion of the same paper[26]:--

"The last discharge of Conybeare's artillery, served by the great Oxford engineer against the Fluvialists, as they are pleased to term us, drew upon them on Friday a sharp volley of musketry from all sides, and such a broadside, at the finale, from Sedgwick as was enough to sink the 'Reliqui?Diluvian?[27] for ever, and make the second volume shy of venturing out to sea."

In a third letter, written to Dr. Fleming, he gives a similar account of the battle between the Diluvialists and Fluvialists, and concludes with these words[28]:--

"I am preparing a general work on the younger epochs of the earth's history, which I hope to be out with next spring. I begin with Sicily, which has almost entirely risen from the sea, to the height of nearly 4,000 feet, since all the present animals existed in the Mediterranean!"

FOOTNOTES:

[9] Now recognised as gault. The identification named above was soon found to be correct.

[10] Life, Letters, and Journals, vol. i. p. 127. Some sentences (for the sake of brevity) are omitted from the quotation.

[11] He was also elected a Fellow of the Royal Society in 1826.

[12] It appeared in the Edin. Journ. Sci., iii. (1825) p. 112, being his first actual publication. Its importance consisted in proving that serpentine was, or rather had been, an igneous rock. If proper attention had been paid to it, fewer mistaken statements and hypotheses would have attained the dignity of appearing in print.

[13] Dr. Gideon A. Mantell, a surgeon by profession, at that time resident in Lewes, who made valuable contributions to the geology of South-East England, and was also distinguished for his popular lectures and books. He died in 1852.

[14] Probably referring to an article on Scrope's "Geology of Central France," in which he shows that he fully accepted the Huttonian doctrine of interpreting the geology of past ages by reference to the causes still at work. It appeared in the Quarterly Review, Oct. 1827, vol. xxxvi. p. 437.

[15] Life, Letters, and Journals, vol. i. p. 169.

[16] Dr. John MacCulloch, author (among other works) of the "Highlands and Western Isles of Scotland." He was an excellent geologist on the mineralogical side, but had little sympathy with paleontology or with the views to which Lyell inclined. He died in 1835.

[17] This district had been already explored by Mr. G. P. Scrope, the first edition of whose classic work, "The Volcanoes of Central France," was published in 1826.

[18] Sir John F. W. Herschel, the second of the illustrious astronomers of that name.

[19] Sir W. J. Hooker.

[20] Certain passages in a letter of Sidonius Apollinaris, Bishop of Clermont, dated about 460 A.D., and in the works of Alcimus Avitus, Archbishop of Vienne, about half a century later, have been interpreted as referring to volcanic eruptions somewhere in Auvergne. This, however, is disputed by many authorities. (See Geological Magazine, 1865, p. 241.)

[21] Life, Letters, and Journals, vol. i. p. 199.

[22] Life, Letters, and Journals, vol. i. p. 238.

[23] Life, Letters, and Journals, vol. i. p. 215.

[24] Life, Letters, and Journals, vol. i. p. 252.

[25] Ibid.

[26] Ut supr? p. 253.

[27] "Reliqui?Diluvian? or Observations on Organic Remains contained in Caves, Fissures, and Diluvial Gravel, and on other Geological Phenomena attesting the Action of an Universal Deluge." By Professor Buckland. 1823.

[28] Ut supr? p. 254.

CHAPTER IV.

THE PURPOSE DEVELOPED AND ACCOMPLISHED.

The summer of 1829 was spent at Kinnordy, when the quarries of Kirriemuir and the neighbouring districts were visited from time to time, the workmen being encouraged to look out for the remains of plants and the scales of fishes. Murchison, however, was again travelling on the Continent, and, in company with Sedgwick, was exploring the geological structure of the Eastern Alps and the basin of the Danube. They appear to have kept up communication with Lyell, who hears with satisfaction of the results of their work, since these cannot fail to keep Murchison sound in the Uniformitarian faith and to complete the conversion of Sedgwick.[29]

"The latter" (Lyell writes to Dr. Fleming) "was astonished at finding what I had satisfied myself of everywhere, that in the more recent tertiary groups great masses of rock, like the different members of our secondaries, are to be found. They call the grand formation in which they have been working sub-Apennine. Vienna falls into it. I suspect it is a shade older, as the sub-Apennines are several shades older than the Sicilian tertiaries. They have discovered an immensely thick conglomerate, 500 feet of compact marble-like limestone, a great thickness of oolite, not distinguishable from Bath oolite, an upper red sand and conglomerate, etc. etc., all members of that group zoologically sub-Apennine. This is glorious news for me.... It chimes in well with making old red transition mountain limestone and coal, and as much more as we can, one epoch, for when Nature sets about building in one place, she makes a great batch there.... All the freshwater, marine, and other groups of the Paris basin are one epoch, at the farthest not more separated than the upper and lower chalk."

A letter to the same correspondent, written nearly three weeks later, at the end of October, and after his return to London, refers to the consequences of this journey.[30]

"Sedgwick and Murchison are just returned, the former full of magnificent views. Throws overboard all the diluvian hypothesis; is vexed he ever lost time about such a complete humbug; says he lost two years by having also started a Wernerian. He says primary rocks are not primary, but, as Hutton supposed, some igneous, some altered secondary. Mica schist in Alps lies over organic remains. No rock in the Alps older than lias.[31] Much of Buckland's dashing paper on Alps wrong. A formation (marine) found at foot of Alps, between Danube and Rhine, thicker than all the English secondaries united. Munich is in it. Its age probably between chalk and our oldest tertiaries. I have this moment received a note from C. Prost by Murchison. He has heard with delight and surprise of their Alpine novelties.

A short time afterwards, in a letter addressed to Mr. Leonard Horner, Lyell declines to become a candidate for the Professorship of Geology and Mineralogy at the London University,[32] which was first opened in the autumn of the previous year. Evidently he considers himself to be too fully occupied, for he writes to Dr. Mantell on December 5th that his book has taken a definite shape.[33] "I am bound hand and foot. In the press on Monday next with my work, which Murray is going to publish--2 vols.--the title, 'Principles of Geology: being an Attempt to Explain the Former Changes of the Earth's Surface by Reference to Causes now in Operation.' The first volume will be quite finished by the end of the month. The second is, in a manner, written, but will require great recasting. I start for Iceland by the end of April, so time is precious." The process of incubation was continued throughout the winter. On February 3rd, 1830, he had corrected the press as far as the eightieth page, getting on slowly, but with satisfaction to himself. "How much more difficult it is," he remarks, "to write for general readers than for the scientific world; yet half our savants think that to write popularly would be a condescension to which they might bend if they would." He fully expects that the publication of his book will bring a hornet's nest about his head, but he has determined that, when the first volume is attacked, he will waste no money on pamphleteering, but will work on steadily at the second volume, and then, if the book is a success, at the second edition, for "controversy is interminable work." He felt now that the facts of nature were on his side, and his conclusions right in the main; so, like most strong men, he adopted the same course as did the founder of Marischal College, Aberdeen, and wrote over the door of his study, "Lat them say."

The plan of a summer tour in Iceland fell through; so did another for a long journey from St. Petersburg by Moscow to the Sea of Azof, to be followed by an examination of the Crimea and the Great Steppe, and a return up the

Danube to Vienna; but by the middle of June the first volume of the "Principles" was nearly finished; and in a letter to Scrope,[34] to whom advance sheets of the book had been forwarded, in order that he might review it in the Quarterly, Lyell explains concisely the position which he has taken in regard to cosmology and the earth's history.

"Probably there was a beginning--it is a metaphysical question, worthy a theologian--probably there will be an end. Species, as you say, have begun and ended--but the analogy is faint and distant. Perhaps it is an analogy, but all I say is, there are, as Hutton said, 'no signs of a beginning, no prospect of an end.' Herschel thought the nebul?became worlds. Davy said in his last book, 'It is always more probable that the new stars become visible, and then invisible, and pre-existed, than that they are created and extinguished.' So I think. All I ask is, that at any given period of the past, don't stop inquiry when puzzled by refuge to a beginning, which is all one with 'another state of nature,' as it appears to me. But there is no harm in your attacking me, provided you point out that it is the proof I deny, not the probability of a beginning. Mark, too, my argument, that we are called upon to say in each case, 'Which is now most probable, my ignorance of all possible effects of existing causes,' or that 'the beginning' is the cause of this puzzling phenomenon?"

In other parts of the letter he refers to his theory of the dependence of the climate of a region upon the geography, not only upon its latitude, but also upon the distribution of land and sea, and that of the coincidence of time between zoological and geographical changes in the past, as the most novel parts of the book; stating also that he has been careful to refer to all authors from whom he has borrowed, and that to Scrope himself he is under more obligation, so far as he knows, than to any other geologist. The concluding words also are interesting:--

"I conceived the idea five or six years ago, that if ever the Mosaic geology could be set down without giving offence, it would be in an historical sketch, and you must abstract mine in order to have as little to say as possible yourself. Let them feel it, and point the moral."

The last-named difficulty, to which Lyell refers in another part of this letter, was undoubtedly one of the most formidable "rocks ahead" in the path of his new book. Up to that time the progress of geology had been most seriously impeded by the supposed necessity of making its results harmonise with the Mosaic cosmogony. It was assumed as an axiom that the opening chapters of Genesis were to be understood in the strict literal sense of the words, and that to admit the possibility of misconceptions or mistakes in matters wholly beyond the cognisance of the writers, was a denial of the inspiration of Scripture, and was rank blasphemy. A large number of persons--among whom are the great mass of amateur theologians, together with some experts--are always very prone to assume the meaning of certain fundamental terms to be exactly that which they desire, and then to proceed deductively to a conclusion as if their questionable postulates were axiomatic truths. They further assume, very commonly, that the possession of theological knowledge--scanty and superficial though it may be--enables them to dispense with any study of science, and to pronounce authoritatively on the value of evidence which they are incapable of weighing, and of conclusions which they are too ignorant to test. Being thus, in their own opinion, infallible, a freedom of expression is, for them, more than permissible, which, in most other matters, would be generally held to transgress the limits of courtesy and to trespass on those of vituperation. Lyell had perceived that little real progress could be made till geologists were free to look facts in the face and to follow their guidance to whatever conclusions these might lead, irrespective of supposed consequences; or that, in other words, questions of science must be settled by inductive reasoning from accurate observations, and not by an appeal to the opinions of the men of olden time, however great might be the sanctity of their characters or the honour due to their memories. Wisely, however, he determined to prefer an indirect to a direct method of attack, and to avoid, so far as was possible, giving needlessly any cause of offence by abruptness of statement or by intemperance of language.

In deluges, the favourite resort of every "catastrophic" geologist, Lyell had long lost faith, and he laughs in one of his letters at the idea of a French geologist, that a sudden upheaval of South America may have been the cause of the Noachian flood. To the breaks in the succession of strata, a fact upon which the catastrophists much relied, he attached comparatively little value, insisting on their more or less local character. In the records of the rocks he finds no trace of a clean sweep of living creatures or of anything like a general clearance of the earth's surface, and no corroboration of the Mosaic cosmogony. He is bent on interpreting the work of Nature in the past by the work of Nature in the present, and not by the writings of the Fathers, or even by the words of Scripture itself.

Some time in the month of June the last sheet of the "Principles" must have been sent to press; for on the 25th of that month Lyell writes from Havre on his way to Bordeaux, through part of Normandy, Brittany, and La Vende. This journey took him, as he says, "through some of the finest countries and most detestable roads he ever saw." On this occasion he was accompanied by a Captain Cooke, a commander in the Royal Navy; a man well informed, acquainted with Spain (the end of their journey), a botanist, and not wholly ignorant of geology--in short, an excellent companion, whose only fault was being "a little too fond of lagging a day for rest," even in places where nothing is to be done. Writing from Bordeaux to a sister, Lyell expresses a hope that at Bagnes de Luchon he may hear whether his book is out.[35] Two passages in his letter are not without a more general interest. One repeats a remark made to him by D'Aubuisson, whom he describes as "a great gun of the old Wernerian school, who ... thinks the interest of the subject greatly destroyed by our new innovation, especially our having almost cut mineralogy and turned it into a zoological science."[36] D'Aubuisson also said, "We Catholic geologists flatter ourselves that we have kept clear of the mixing of things sacred and profane, but the three great Protestants, De Luc, Cuvier, and Buckland, have not done so; have they done good to science or to religion? No, but some say they have to themselves by it." The other remark is interesting in its reference to French politics, seeing that it is dated on the 9th of July, 1830. It runs thus[37]:--

"The quiet and perfect order and calmness that reigned at Bourbon and Bordeaux and Toulouse during the heat of the elections, afford a noble example to us--never were people in a greater state of excitement on political grounds than the French at this moment, yet never in our country towns were Assizes conducted with more seriousness and quiet. There is no occasion to make the rabble drunk. All the voters of the little colleges are of the rank of shopkeepers at least, those of the highest are gentlemen--only 20,000 of them out of the 30 millions of French. They are too many for such jobbing as in a Scotch county, and too independent and rich to have the feelings of a mob."

Yet at the end of this month came the "three days of July"; "perfect order and calmness" were at an end; Charles X. abdicated the throne, and the Bourbons again became exiles from France.

From Toulouse Lyell and his companion journeyed by the banks of the Arie to the picturesque old town of Foix, and from this place to Ax, a watering-place on one of the tributaries to that river, in the heart of the Pyrenees. His keen eye notes at once the difference between the scenery of this chain and that of the Alps. Apart from the different character of the vegetation--the more luxuriant flora, the extensive forests of beech and oak at elevations where in Switzerland only the pines and larches would flourish--the valleys are narrower, the mountains more precipitous--the scenery, in short, is more like that around Interlaken or in the valley of Lauterbrunnen, without the lakes of the one or the grand background of snowy peaks in the other. In the Pyrenees the inferior height and the more southern position of the chain diminishes the snowfields and curtails the glaciers, so that the torrents run with purer waters, like they do in the Alps about the birthplace of the Po.

In order to acquire a clear idea of the structure of the Pyrenees the travellers crossed from Ax to the southern side of the watershed, though they still remained on French territory; for here, in the neighbourhood of Andorre, the frontier cuts off the heads of one or two valleys which geographically form part of Spain. Into this country they had purposed to descend, but the obstacles interposed by the reactionary jealousy of local Dogberries and the possible risks from political complications were so great, that they judged it wiser to abandon the attempt. So the travellers separated for a time, Captain Cooke, who feared the heat of the lower country, going eastwards through the curious little mountain republic of Andorre to Luchon; while Lyell, who seems to have been proof against the sun, recrossed the watershed into the valley of the Tet and descended it to Perpignan. Information obtained in this town encouraged him to go direct to Barcelona, where the Captain-General, the Conde D'Espagne, a distinguished soldier and diplomatist, gave him a courteous reception, and did everything in his power to smooth the way for a visit to Olot, a region of extinct volcanoes, which had been one of the chief ends of Lyell's journey. The expedition was successful; he did not fall among thieves, and was only annoyed by the tedious formalities and petty impertinences of the local functionaries of northern Spain; and he returned to France by a pass on the eastern side of the Canigou. He was not a little astonished, as might be expected from the remarks already quoted, when he found on arriving in that country that the reign of the Bourbons and the priests was over, the tricolor flag was hoisted on all the churches, and the royalist officials had been replaced by the nominees of the National Government.

The visit to Olot amply repaid him for the toil and trouble of the journey. An account of the district was inserted in the concluding volume of the "Principles," which was afterwards incorporated into the "Elements of Geology." The following summary is quoted from a letter to Scrope, who had suggested the visit, which was written from Luchon, where he arrived a few days after his return into France[38]:--

"Like those of the Vivarais [the volcanoes of Catalonia] are all, both cones and craters, subsequent to the existence of the actual hills and dales, or, in other words, no alteration of previously existing levels accompanied or has followed the introduction of the volcanic matter, except such as the matter erupted necessarily occasioned. The cones, at least fourteen of them mostly with craters, stand like Monpezat, and as perfect; the currents flow down where the rivers would be if not displaced. But here, as in the Vivarais, deep sections have been cut through the lava by streams much smaller in general, and at certain points the lava is fairly cut through, and even in two or three cases the subjacent rock. Thus at Castel Follet, a great current near its termination is cut through, and eighty or ninety feet of columnar basalt laid open, resting on an old alluvium, not containing volcanic pebbles; and below that, nummulitic limestone is eroded to the depth of twenty-five feet, the river now being about thirty-five feet lower than when the lava flowed, though most of the old valley is still occupied by the lava current. There are about fourteen or perhaps twenty points of eruption without craters. In all cases they burst through secondary limestone and sandstone, no altered rocks thrown up, as far as I could learn, not a dike exposed. A linear direction in the cones and points of eruption from north to south. Until some remains of quadrupeds are found, or other organic medals found, no guess can be made as to their geological date, unless anyone will undertake to say when the valleys of that district were excavated. As to historical dates, that is all a fudge ... I can assure you that there never was an eruption within memory of man."

At Luchon Lyell rejoined Captain Cooke, and they visited one or two interesting spots in the more western part of the Pyrenees, such as the Cirque de Gavarnie and the Brahe de Roland. The former would afford object-lessons on the erosive action of cascades; the latter would set him speculating on the causes which could have fashioned that strange portal in the limestone crest of the mountain. They descended some distance on the Spanish side of the Brahe, in order to make a more complete investigation of the structure of the chain, sleeping at a shepherd's hut and returning across the snowfields next day. It is evident that whenever there was a hope of securing any geological information or of seeing some remarkable aspect of nature, Lyell was almost insensible either to heat or to fatigue.

Towards the middle of September he had reached Bayonne, from which place another very interesting letter is despatched to Scrope.[39] In this he gives suggestions for making a number of experiments in order to produce by artificial means such rock-structures as lamination, ripple-mark, and current-bedding, and describes briefly a series of observations bearing on these questions, which had been carried out both during his late journey and on other occasions. "I have," he says, "for a long time been making minute drawings of the lamination and stratification of beds, in formations of very different ages, first with a view to prove to demonstration that at every epoch the same identical causes were in operation. I was next led in Scotland to a suspicion, since confirmed, that all the minute regularities and irregularities of stratification and lamination were preserved in primary clay-slate, mica-slate, gneiss, etc., showing that they had been subjected to the same general and even accidental circumstances attending the sedimentary accumulation of secondary and fossil-bearing formations.[40] Lastly, I came to find out that all these various characters were identical with those presented by the bars, deltas, etc., of existing rivers, estuaries, etc."

Early in October Lyell is back again in Paris, to find Louis Philippe seated on the throne in the place of Charles X., and a war party "praying night and day for the entry of the Prussians into Belgium in the hope of the French being drawn into the affair. A finer opportunity, they say, could not have happened for resuming our natural limits on the Rhine." In the midst of political changes and warlike aspirations geology, he observes, is not making much progress in Paris. Some of the naturalists have "got their heads too full of politics"; others are forced to work as literary hacks in order to live. "Books on natural history and medicine have no sale; there is a demand only for political pamphlets."

So Lyell enters into an engagement with Deshayes, who, like so many others, has to live by his pen lest he should starve by science, for "a private course of fossil conchology," and for two months' work after Lyell has returned to England, to be spent in tabulating the species of Tertiary shells in his own (Deshayes') and the other great collections of Paris. "I shall thus," Lyell says, "be giving the subject a decided push by rendering the greater wealth of the French collectors available in illustrating the greater experience of the English geologists in actual observation; for here they sit still and buy shells, and work indoors, as much as we travel." He also remarks to the same correspondent (a sister): "I am nearly sure now that my grand theory of temperature will carry the day.... I will treat our geologists with a theory for the newer deposits in next volume, which, although not half so original, will perhaps surprise them more."[41] He was expecting, as another letter shows, to prove the gradual approximation of the fauna preserved in the Tertiary deposits to that which still exists, and to settle, as he hopes "for ever, the question whether species come in all at a batch or are always going out and coming in." Already he is in a position to affirm that the Tertiary formations of Sicily in all probability are more recent than the "crags" of England, for, among the sixty-three species which he had collected from the beds underlying Etna, only three were not known to be still inhabitants of the Mediterranean; and besides this, between these "crags" and the London clay a series of formations can be intercalated. In the same letter (to Scrope)[42] he states that Deshayes has found, at St. Mihiel on the Meuse, three old needles of limestone, like those in the Isle of Wight, round which run three distinct lines of perforations, like those on the columns of the "Temple of Serapis;" these hollows being "sometimes empty, but thousands of them filled with saxicavas." This, of course, was a proof that there had been, in comparatively recent times, important changes in the level of the land and sea.

Early in November Lyell is back in London, at his chambers in Crown Office Row, Temple, to find that Scrope's review of the first volume of the "Principles" has been much admired, that the book is selling steadily, and is likely to prove "as good as an annuity"; that it has not been seriously attacked by the "Diluvialists," while it has been highly praised by the bulk of geologists. He is about to move, he writes, into chambers in Raymond Buildings, Gray's Inn, which are "very light, healthy and good, on the same staircase as Broderip." Invitations to dinner are becoming frequent, but he wisely determines to go but little into society. "All my friends," he says, "who are in practice do this all the year and every year, and I do not see why I should not be privileged, now that I have the moral certainty of earning a small but honourable independence if I labour as hard for the next ten years as during the last three. I was never in better health, rarely so good, and after so long a fallow I feel that a good crop will be yielded and that I am in good train for composition."[43] The second volume, he hopes, will be out in six months; this will include the history of the globe to the beginning of the Tertiary era, when the first of existing species appeared.

The next year, 1831, was an epoch marked by more than one change. To take the smallest first, he was made a deputy-lieutenant of the county of Forfar; next, in March, he was elected Professor of Geology at King's College, London, which had been recently founded by members of the Church of England as an educational counterpoise to the University of London (University College). To Lyell himself the appointment was comparatively unimportant, but it indicated that wider views on scientific questions and a more tolerant spirit were gaining ground among the higher ranks of the clergy in the Established Church. The appointment was in the hands, exclusively, of the Archbishop of Canterbury, the Bishops of London and of Llandaff, and two "strictly orthodox doctors." Llandaff, Lyell was informed, hesitated, but Conybeare,[44] though opposed to Lyell's theories, vouched for his orthodoxy. So the prelates declared that they "considered some of my doctrines startling enough, but could not find that they were come by otherwise than in a straight-forward manner, and (as I appeared to think) logically deducible from the facts; so that whether the facts were true or not, or my conclusions logical or otherwise, there was no reason to infer that I had made my theory from any hostile feeling towards revelation"[45]--a conclusion, marked by a wise caution, which representatives of the Church of England would have done well to bear in mind on more than one subsequent occasion--such as, for example, when the question of the antiquity of man or that of the origin of species was raised. But supporters of the Church of England may fairly maintain that in difficult crises, especially in those connected with discoveries in science or in history, the utterances of her bishops have been generally cautious and far-seeing; displays of confident ignorance and rash denunciations are more common among the "inferior clergy." As a comment on the moderation indicated by his election, Lyell says that a friend in the United States affirms that there "he could hardly dare to approve of the doctrines even in a review, such a storm would the orthodox raise against him. So much for toleration of Church Establishment and No Church Establishment countries." A third event of the year--which also happened in the earlier part of it--was destined to exercise a much more lasting influence upon his life. This was his engagement to Miss Mary Horner, eldest daughter of Mr. Leonard Horner, the younger and hardly less distinguished brother of Francis Horner, who, while almost as enthusiastic a geologist as his future son-in-law, took an active interest in educational questions, and afterwards did public service as Inspector of Factories.

By the middle of June Lyell had advanced as far as page 110 in printing the second volume of the "Principles of Geology," notwithstanding interruptions, such as a visit to Cambridge, where he took an ad eundem degree,[46] and the presence of his father and brother, as well as of his friend Conybeare, in London, all of whom required to be lionised. The letter[47] (to Mantell) which refers to these impediments, passes abruptly from Fitton's broken arm to the giant femur of a new reptile, and incidentally mentions the discovery of a section which has since become a centre of geological controversy. "Murchison and his wife," he writes, "are gone to make a tour in Wales, where a certain Trimmer has found near Snowdon 'crag' shells at a height of 1,000 feet, which Buckland and he convey thither by the deluge." The shells are at an altitude above sea-level considerably higher than Lyell supposed. Moel Tryfaen is a massive, rather outlying hill, about five miles west of the peak of Snowdon, and at about the same distance from the nearest part of the sea-coast. Its bare summit rises gently to a scattered group of projecting crags, the highest of which is 1,401 feet above the sea. On the eastern side are extensive slate quarries, and in working these the shell beds are disclosed a short distance below the summit. They consist of well-stratified sands, with occasional gravelly beds, and contain a fair number of shells, both broken and whole, the fauna being slightly more arctic than that which still inhabits the neighbouring sea. The deposit is now recognised as more recent than the "crags" of East Anglia, for none of the species are extinct, and is assigned to some part of the so-called Glacial Epoch. It was before long regarded as an indication that, at no very remote date after North Wales had assumed or very nearly assumed its present outlines, the whole district was depressed for at least 1,380 feet, so that the sea broke over the summit crags of Moel Tryfaen. For many years this interpretation passed unquestioned; but a modern school of geologists has found it to be such an inconvenient obstacle to certain hypotheses about the former extent of land-ice, that they maintain these shells were collected from the bed of the Irish Sea (then supposed to be above water) by an ice-sheet as it was on its way from the north to invade the Principality, and were conveyed by it, with all care, up the slopes of Moel Tryfaen, till they were finally deposited on its summit, in beds which somehow or other were stratified. One may venture to doubt whether the hypothesis of a rampant and conchologically-disposed ice-sheet would have found much more favour with the cautiously inductive mind of Lyell than that of a deluge.

Shortly after this letter, Lyell, though all the manuscript of his second volume had not yet been sent to the printers, and proof-sheets followed him, refreshed himself with a tour of four or five weeks in the volcanic district of the Eifel. Here the cones, all comparatively low, are scattered sporadically over a rolling upland which occupies the angle between the Rhine and the Moselle. The valleys for the most part are carved out of slaty rocks much of the same age as those of Devonshire; and the craters, "strange holes, each eruption having been almost invariably at some new point," are now very commonly occupied by quiet pools of water, such as Lyell had already seen in the old volcanic districts of the Papal States. Among these craters, composed sometimes of loose and light scoria, from which no lava-stream ever flowed, he found fresh evidence--as at the Rotherberg--against the diluvian hypothesis. "It is," as he writes to his friend, Dr. Fleming, "one of the ten thousand proofs of the incubus that the Mosaic deluge has been, and is, I fear, long destined to be, on our science. Now, I am fully determined to open my strongest fire against the new diluvial theory of swamping our continents by waves raised by paroxysmal earthquakes. I can prove by reference to cones (hundreds of uninjured cones) of loose volcanic scori?and ashes, of various and some of great antiquity (as proved by associated organic remains), that no such general waves have swept over Europe during the Tertiary era--cones at almost every height, from near the sea, to thousands of feet above it."[48]

But early in August he was back in London, hard at work in writing and correcting proofs. This business detained him longer than he anticipated, but his labours were cheered by the news of the eruption of Graham's Island. Here was another case in support of the thesis which he was ready to maintain against all comers. But a few months since there had been a depth of eighty fathoms, as was proved by sounding, on the site of this island. Now the cone "is 200 feet above water and is still growing.[49] Here is a hill 680 feet, with hope of more, and the probability of much having been done before the 'Britannia' sounded." Surely Nature herself was testifying "her approbation of the advocates of modern causes! Was the cross which Constantine saw in the heavens a more clear indication of the approaching conversion of a wavering world?"

But in the beginning of September Lyell broke away from the emissaries of the press and took passage by sea to Edinburgh, there to combine business with a fair amount of both scientific work and social pleasure. This visit afforded him an opportunity of hearing Chalmers preach. In a letter to Miss Horner he gives a brief abstract, and expresses his general opinion of the sermon[50]:--

"It was a very long discourse, but admirable. The subject was 'repentance,' a hackneyed one enough.... He explained the effect of habit, and its increasing power over the mind, as a law of our nature, with as much clearness and as philosophically as he could have done had he been explaining the doctrine to a class of university students in a lecture on the philosophy of the human mind. But then the practical application was enforced by a strain of real eloquence, of a very energetic, natural, and striking description.... But, unfortunately, every here and there he seemed to feel that he was sinning against some of the Calvinistic doctrines of his school, and all at once there was some dexterous pleading about 'original sin,' which interfered a little with the free current of the discourse.... Upon the whole, however, judging from this single specimen, I think I would sooner hear him again than any preacher I ever heard, Reginald Heber not excepted."

At this time Lyell was keeping a journal, which was forwarded to Miss Horner, then in Germany, to serve apparently as a substitute for ordinary letters; home news, disturbances arising from the struggle over the Reform Bill, visits of friends, geological researches, walks on the hills to search for plants or for insects, the habits of the Kinnordy bees, or the accomplishments of two parrots, brought from Africa by his naval brother--all being jotted down just as they occurred.

Among this farrago--though not of nonsense--geological topics, since Miss Horner had similar tastes, occupy a considerable space. She, however, evidently was, comparatively speaking, a beginner, and in one or two characteristic sentences her lover and preceptor passes from information to counsel: "If you are not frightened by De la Beche, I think you are in a fair way to be a geologist; though it is in the field only that a person can really get to like the stiff part of it. Not that there is really anything in it that is not very easy, when put into plainer language than scientific writers choose often unnecessarily to employ." He also records[51] a piece of advice from his old friend, Dr. Fleming, which is enough to make a modern professor of geology sigh for "the good old times." He said to Lyell:

"If you lecture once a year for a short course, I am sure you will derive advantage from it. A short practice of lecturing is a rehearsal of what you may afterwards publish, and teaches you by the contact with pupils how to instruct, and in what you are obscure. A little of this will improve your power, perhaps as an author. Then, as you are pursuing a path of original and purely independent discovery and observation, it increases much your public usefulness in a science so unavoidably controversial to have thrown over you the moral protection of being in a public and responsible situation, connected with a body like King's College. But then you must stipulate that you are to be free to travel, and must only be bound to give one short course annually."

Truly those must have been halcyon days for professors!

The journal also proves, by its brief account of a Scotch festival, which accords with little hints dropped elsewhere in it or in letters, that our forefathers, not wholly excluding men of science, some sixty years ago habitually consumed much more "strong drink" than would be considered correct at the present day:--

"It was just an Angus set-to of the old regime. They arrived at half-past six o'clock and waited dinner one hour. Gentlemen rejoined the ladies at half-past twelve o'clock! They, in the meantime, had had tea, and a regular supper laid out in the drawing-room. After an hour with the ladies they returned to the dining-room to supper at half-past one o'clock, and my father left them at half-past two o'clock! The ladies did not go to this supper."

The journal, in short, like the well-known Scotch dish, affords a great deal of "confused feeding" of a pleasant sort, but no samples of love-making. The nearest approach to it is in the following passage, which is worth quoting, not for that reason, but as incidentally disclosing the strength of the author's character:--

"I shall write a few words before I get into the steamboat just to tranquillise my mind a little, after reading several controversial articles by Elie de Beaumont and others against my system. If I find myself growing too warm or annoyed at such hostile demonstrations I shall always retreat to you. You will be my harbour of peace to retire to, and where I may forget the storm. I know that by persevering steadily I shall some years hence stand very differently from where I now am in science; and my only danger is the being impatient, and tempted to waste my time on petty controversies and quarrels about the priority of the discovery of this or that fact or theory."[52]

Friends in plenty were awaiting him in London, which was reached about the first of November: the Murchisons and Somervilles, Broderip, Curtis, Basil Hall, and Hooker, with Necker from Switzerland, and many more. He is also cheered by finding that his ideas are steadily gaining ground among geologists, converts becoming more confident, unbelievers more uneasy. He made good progress with his book, and realised, before the end of the year, that his materials could not be compressed into a single volume; so he determined to issue the part already completed as a second volume, and to finish the work in a third.

From time to time the diary contains references to a recent contest for the Presidency of the Royal Society, and to political matters such as the Reform Bill; but, though in favour of the latter, he is not very enthusiastic on the subject, for on one occasion he expresses regret at having been absent, through forgetfulness, from a meeting of the Geographical Society, where he would have "got some sound information instead of hearing politicians discuss the interminable bill."

The lectures at King's College evidently weighed upon his mind as they drew near, and he was not stirred to enthusiasm by the prospect of teaching; for towards the close of the year he more than once debated with his friends the question whether or no he should retain the appointment. Murchison was in favour of resignation; Conybeare took the opposite view. Of his advice Lyell remarks, "The fact is, Conybeare's notion of these things is what the English public have not yet come up to, which, if they had, the geological professorship in London would be a worthy aim for any man's ambition, whereas it is now one that the multitude would rather wonder at one's accepting."[53] The British public apparently still lags a long way behind the Conybearian ideal, and retains its contempt for all those who, by presuming to teach, insinuate doubts as to its innate omniscience.

Lyell, however, clearly perceived that it was absolutely necessary that every teacher of professorial rank should be himself a pioneer in his subject--a fact of which government officials, as a rule, seem to be totally ignorant. His comments, a little later in the year, on the arrangements at the University of Bonn are worth recording. "The Professors have to lecture for nine months in the year--too much, I should think, for allowing time for due advancement of the teacher." Lyell's desires in regard to remuneration seem reasonable enough. He is anxious to earn by his scientific work enough to provide for the extra expenses which this work entails, and yet to command sufficient time to advance his knowledge and reputation. The fates proved more propitious to him than they are generally to men of science, for he succeeded in accomplishing both of his desires.

Little of importance happened during the early part of 1832. There was plenty of hard work in collecting facts, in consulting friends about special difficulties, and in working at the manuscript for the third volume of the "Principles," for the second made its appearance almost with the new year. Toil was sweetened by occasional pleasures, such as an evening with the Somervilles, or a dinner party at the Murchisons, a talk with Babbage or Fitton, or a symposium at the Geological Club, at which it is sometimes evident that good care was taken lest science should become too dry. One passage in his diary indicates that sixty years have considerably changed the habits of life in town and in the country, for at the present day most people would express themselves in the opposite sense. "I have enjoyed parties and two plays this month very much, because it was recreation stolen from work; but the difficulty in the country is that, on the contrary, one's hours of work are stolen from dissipation."

The lectures at King's College were begun in May. Lyell evidently was not a nervous man, but he regarded the near approach of this new kind of work with some trepidation, and admits that he slept ill before the first lecture. It was, however, a decided success in every respect, and the audience was a large one, for the Council, after some hesitation, had permitted the attendance of ladies. Each lecture was pronounced by the hearers to be better than the last, and Lyell uses the opportunity, as he says, to fire occasional shots at Buckland, Sedgwick, and others who are still hankering after catastrophic convulsions and all-but universal deluges. As a further encouragement, his publisher, Murray, agrees willingly to a reprint of the first volume of the "Principles," and only hesitates between an edition of 750 or of 1,000 copies. About this time, also, he was asked to undertake the presidency of the Geological Society, but that, notwithstanding Murchison's urgency, he firmly declined for the present; writing of it to Miss Horner, "It is just one of those temptations the resisting of which decides whether a man shall really rise high or not in science. For two more years I am free from les affaires administratives, which, said old Brochart in his late letter to me, have prevented me from studying geology d'une manie suivie, whereby you have already carried it so far."

He was, however, soon to be engrossed in an "affair" of another kind; one which has proved very detrimental to the progress of many men of science, but which, in Lyell's case, had the happiest results, and smoothed rather than it impeded his path to fame; for in the summer--on July 12th--he ceased to be a bachelor. The marriage was celebrated at Bonn, where Miss Horner's family were still resident. A Lutheran clergyman seems to have officiated, and the ceremony was a very quiet one; the distance from home preventing the attendance of English friends or even of relations of the bridegroom.

The newly-married couple departed from Bonn up the Rhine, and travelled by successive stages to Heidelberg, but they were not forgetful of geology, even in the first week of the honeymoon, for they visited as they journeyed more than one interesting section on the western edge of the Odenwald. Then they made excursions to Carlsruhe and Baden-Baden, and ultimately travelled from Freiburg to Schaffhausen through the romantic defiles of the Helenthal, and across the corner of the Black Forest. A journal was now needless, and probably the newly-married couple were too much engrossed with their own happiness to write many letters, for few details have been preserved about their Swiss tour. It was, however, comparatively a short one, for they remained less than a fortnight in the country. Still Lyell probably found it useful in refreshing recollections and testing his early impressions by greatly increased knowledge and experience. From the valley of the Rhone they crossed the Simplon Pass into Italy and followed the usual road to Milan along the shore of the Lago Maggiore.

How long they remained in Italy, or by what route they returned to England, is not stated; indeed, for nearly six months next to nothing is on record concerning Lyell's movements or work, but in the beginning of 1833 he and his wife were settled in London at No. 16, Hart Street, Bloomsbury, which became their residence for some years. A state of happiness is not always indicated by much correspondence: probably it was so with Lyell; at any rate, a single letter, dated January 5th, gives the only information of his doings between September, 1832, and April, 1833. In this letter, however, he mentions that the Council of King's College had decided that in future ladies should not be admitted to Lyell's lectures, and that, in consequence, he had received a pressing invitation from the managers of the Royal Institution to give, after Easter, a course of six or eight lectures in their theatre, coupled with the offer of a substantial remuneration.

At the end of April, as he tells his old friend Mantell, both these courses had been begun. The one at the Royal Institution was attended by an audience of about 250, that at King's College, after the opening lecture, dropped down to a class of fifteen. The falling-off was entirely due to the above-named resolution. For this the Council had assigned a reason, which, perhaps, was not a prudent course, for bodies of that kind, when they give reasons, often succeed only in "giving themselves away." The presence of ladies was forbidden, "because it diverted the attention of the young students, of whom," Lyell remarks sarcastically, "I had two in number from the college last year and two this." Had the Council stated boldly that the College did not appoint professors to lecture urbi et orbi, their policy, though it would have appeared a little selfish and might have proved shortsighted, would have been defensible, because the institution was founded for the education of a particular class. But the reason assigned was open to Lyell's retort, and gave the impression of unreality. It is not impossible that the decision was the result of secret "wire-pulling," and represented not so much a fear of the disturbing influence of the fair sex as a dread of the popularity of the subject. Geology was still regarded with grave distrust by a very large number of people, and King's College, it must be remembered, was founded in the supposed interests of the Church of England and in the hope of neutralising the effects of the unsectarian institution in Gower Street. Many of its supporters may have been characterised rather by the ardour of their dislikes than by the width of their sympathies, and may have put pressure on the Council, so that this body may have considered it safer to risk driving a popular man from their staff than to alienate an important section of their adherents and to expose the College to the danger of being charged with lending itself to heretical teaching.[54]

The preparation of these lectures must have been attended with some difficulty, for Lyell writes that, "like all the world," he and his household--everyone except his wife--had been down with the influenza, which in that year was even more rampant in London than it has been in any of its recent visits. But, notwithstanding this and any other interruptions, the third and final volume of the "Principles of Geology" made its appearance in the month of May, 1833.

FOOTNOTES:

[29] Life, Letters, and Journals, vol. i. p. 255.

[30] Ut supr? p. 256.

[31] Further work has not verified some of these statements. There can be no question that a great deal of rock in the Alps is much older than even the Trias. The apparent superposition of crystalline schists to rocks with fossils is due to over-folding or over-thrust faulting--i.e. the schists are the older rocks. Though the Secondary rocks of the Alps have undergone, in places, some modification and mineral changes, these are very different from the metamorphism of those crystalline schists which have a stratified origin.

[32] Now "University College," London, having been incorporated by Royal Charter under that title in November, 1836.

[33] Ut supr? p. 258.

[34] Life, Letters, and Journals, vol. i. pp. 269-271.

[35] When he left the publisher had not decided whether it should be issued

at once or kept back till October.

[36] D'Aubuisson, as time has shown, foresaw a real danger. The neglect of, if not contempt for, mineralogy, which became conspicuous between the years 1840 and 1870, or thereabouts, seriously impeded the progress of geology, at any rate in England.

[37] Life, Letters, and Journals, vol. i. p. 276.

[38] Life, Letters, and Journals, vol. i. p. 283.

[39] Life, Letters, and Journals, vol. i. p. 296.

[40] Subsequent experience has shown that, while the above observations are beyond all question in the case of ordinary sedimentary rocks, structures curiously resembling lamination and ripple-mark may be produced in certain gneisses and crystalline schists by other causes. Still, in many schists, they have originated in the way suggested by Lyell, and indicate that the rock formerly was deposited by water.

[41] Life, Letters, and Journals, vol. i. p. 303.

[42] Ut supr? p. 305.

[43] Life, Letters, and Journals, vol. i. p. 313.

[44] The Rev. W. D. Conybeare, afterwards Dean of Llandaff, an eminent geologist, rather senior to Lyell.

[45] Life, Letters, and Journals, vol. i. p. 316.

[46] It was formerly conceded by the Universities of Oxford, Cambridge, and Dublin that a Master of Arts in any one could assume, under certain conditions, the same position in the others. This carried with it some privileges, though not the suffrage and the full rights of the degree. Lyell had proceeded to the degree of M.A. at Oxford in 1821.

[47] Life, Letters, and Journals, vol. i. p. 318.

[48] Life, Letters, and Journals, vol. i. p. 328.

[49] Ut supr? p. 329. By the end of October it had not only ceased to grow, but also had been nearly washed away by the sea. Now its position is marked by a shoal.

[50] Life, Letters, and Journals, vol. i. p. 331.

[51] Life, Letters, and Journals, vol. i. p. 342.

[52] Life, Letters, and Journals, vol. i. p. 347.

[53] Life, Letters, and Journals, vol. i. p. 358.

[54] Lyell resigned the Professorship after he had finished the course.

CHAPTER V.

THE HISTORY AND PLACE IN SCIENCE OF THE "PRINCIPLES OF GEOLOGY."

The publication of the last volume of the "Principles of Geology" formed an important epoch in Lyell's life. It brought to a successful close a work on which his energies had been definitely concentrated for nearly five years, and for which he had been preparing himself during a considerably longer time. It placed him, before his fourth decade was completed, at once and beyond all question in the front rank of British geologists; it carried his reputation to every country where that science was cultivated. It proved the writer to be not only a careful observer and a reasoner of exceptional inductive power, but also a man of general culture and a master of his mother tongue. The book, moreover, marked an epoch in geology not less important; it produced an influence on the science greater and more permanent than any work which had been previously written, or has since appeared--greater even than the famous "Origin of Species by Natural Selection," for that dealt only with one portion of geology--viz. with pal盤ntology, while the method of the Principles affected the science in every part. For a brief interval, then, we may desert the biography of the author for that of the book--the parent for his offspring--and call attention to one or two topics which are more immediately connected with the book itself. A brief sketch of its future history may be placed first; for, as its author was constantly labouring to improve and perfect his work, it underwent many changes in form and arrangement during the remainder--some two-and-forty years--of his life, which will be better understood from a connected statement than if they have to be gathered from scattered references in the other chapters of his biography.

The first volume of the "Principles of Geology" appeared, as has been mentioned, in January, 1830; the second in January, 1832; and the third in May, 1833. But a second edition of the first volume was issued in January, 1832, and one of the second volume in the same month of 1833; these were all in 8vo size. A new edition of the whole work was published in May, 1834. This, however, took the form of four volumes 12mo. This edition was called the third, because the first two volumes of the original work had gone through second editions. A fourth edition followed in June, 1835, and a fifth in March, 1837.

Thus far the "Principles" continued without any substantial alteration, but the author made an important change in preparing the next edition. He detached from it the latter part--practically, the matter comprised in the third volume of the original work. This he rewrote and published separately as a single volume in July, 1838, under the title of "Elements of Geology"; a sixth edition of the "Principles," thus curtailed, appeared in three volumes 12mo, in June, 1840. The effect of the change was to restrict the "Principles" mainly to the physical side of geology--to the subjects connected with the morphological changes which the earth and its inhabitants alike undergo. Thus it made the contents of the book accord more strictly with its title, while the "Elements" indicated the working out of the aforesaid principles in the past history of the earth and its inhabitants--that is, the latter book deals with the classification of rocks and fossils, or with petrology and historical geology. The subsequent history of the "Elements" may be left for the present.

In February, 1847, the seventh edition of the "Principles" appeared, in which another change was made. This, however, was in form rather than in substance, for the book was now issued in a single thick 8vo volume. The eighth edition, published in May, 1850; and the ninth, in June, 1853, followed the same pattern. A longer interval elapsed before the appearance of the tenth edition, and this was published in two volumes, the first being issued in November, 1866, and the second in 1868. In this interval--more than thirteen years--the science had made rapid progress, and the process of revision had been in consequence more than usually searching. The author, as he states in the preface, had "found it necessary entirely to rewrite some chapters, and recast others, and to modify or omit some passages given in former editions." Many new instances were given to illustrate the effect which forces still at work had produced upon the earth's crust, and these strengthened the evidence which had been already advanced. Into the accounts of Vesuvius and Etna much important matter was introduced, the result of visits which, as we shall find, Lyell made in 1857 and 1858; the chapters relating to the vicissitudes of climate in past geological ages were entirely rewritten, together with that discussing the connection between climate and the geography of the earth's surface; and a chapter, practically new, was inserted, which considered "how far former vicissitudes in climate may have been influenced by astronomical changes; such as variations in the eccentricity of the earth's orbit, changes in the obliquity of the ecliptic, and different phases of the precession of the equinoxes." But the most important change was made in the later part of the book--the last fifteen chapters.[55] These either were entirely new, or presented the original material in a new aspect. In the earlier editions of his work, Lyell had expressed himself dissatisfied, as we have already seen, with the idea of the derivation of species from antecedent forms by some process of modification, and had pointed out the weak places in the arguments which were advanced in its favour. But the evidence adduced by Darwin and Wallace in regard to the origin of species by natural selection, strengthened by the support of Hooker on the botanical side, had removed the difficulties which the cruder statements of Lamarck and other predecessors had suggested to his mind, so that Lyell now appears as a convinced evolutionist. The question also of the antiquity of man is much more fully discussed than it had been in the earlier editions.

Considerable changes were introduced into the eleventh edition, which appeared in January, 1872, but these were chiefly additions which were made possible by the rapidly increasing store of knowledge, as, for instance, much important information concerning the deeper parts of the ocean. On this interesting subject great light had been thrown by the cruises of the several exploring vessels, notably those of the Lightning, the Bulldog, and the Porcupine, commissioned by the British Government--cruises in the course of which soundings had been taken and temperatures observed in the North Atlantic down to depths of about 2,500 fathoms; and in the lowest parts of the western basin of the Mediterranean. Samples also of the bottom had been obtained, and, in many cases, even dredgers had been successfully employed at these depths. Thanks to the skill of the mechanician, the way had been opened which led into a new fairyland of science. This was not, like some fabled Paradise, guarded by mountain fastnesses and precipitous ramparts of eternal snow; it was not encircled by storm-swept deserts, or secluded in the furthest recesses of forests, hitherto impenetrable; but it lay deep in the silent abysses of ocean--on those vast plains, which are unruffled by the most furious gale, or by the wildest waves. In these depths, beneath the tremendous pressure of so vast a thickness of water, and far below the limits at which the existence of life had been supposed to be possible, numbers of creatures had been discovered--many of them strange and novel: molluscs, sea-lilies, glassy sponges of unusual beauty--creatures often of ancient aspect, relics of a fauna elsewhere extinct; and the ocean floor, on and above which they moved, was strewn with the white dust of countless coverings of tiny foraminifera, which, even if none were actually living, had fallen like a gentle but incessant rain from the overhanging mass of water.

Similar changes were introduced into the twelfth edition of the "Principles," upon which the author was engaged even up to the last few weeks of his life. The Challenger, it will be remembered, started on her memorable voyage of exploration at the close of the year in which the eleventh edition had appeared; and though she did not actually return till after Lyell's death, notes of some of her most interesting discoveries had been communicated from time to time to the scientific journals of this country. The edition, however, was left incomplete. The first volume had been passed for the press, but the second was still unfinished; so that this twelfth edition was posthumous, the work of revision having been finished by the author's nephew and heir, Mr. Leonard Lyell.

By such conscientious and unremitting labour, the scientific value of the "Principles" was immensely increased; it kept always in step with the advance of the science, but at the same time it lost, as was inevitable, a little of that literary charm and that sense of freshness which was at first so marked a characteristic. Books, like children, are apt to lose some of their beauty as they increase in size and strength. One must compare an early and a late edition, such as the first or third and the tenth or eleventh, in order to realise how great were the changes in this passage from childhood to adolescence. New material was incorporated into every part; it makes its appearance sometimes on every page; changes are made in the order of the subjects; many chapters are entirely rewritten; nevertheless, a considerable portion corresponds almost word for word in the two editions. Lyell was no hurried writer, or "scamper" of work; he paid great attention to composition, so that when the facts which he desired to cite had undergone no change, he very seldom found any to make in his language. Nevertheless, here and there, some small modification, a slight verbal difference, a trifling alteration in the order of a sentence, the insertion of a short clause to secure greater perspicuity, shows to how careful and close a revision the whole had been subjected. In the substance of the work, besides the excision of nearly one-third of the material and the complete reconstruction of the part relating to the antiquity of man and the origin of species, already mentioned, the following are the most important changes. The chapters which discuss the evidence in favour of past mutations of climate and the causes to which these are due, are rewritten and greatly enlarged. In the earlier editions, the effects of geographical changes were regarded as sufficient to account for all the climatal variations that geology requires; in the later editions, the possible co-operation of astronomical changes is admitted. Great additions also are made to the parts referring to the condition of the bed of the ocean, and much new and important information is incorporated into the sections dealing with volcanoes and earthquakes; including many valuable observations which had been made during visits to Vesuvius and to Etna in the autumns of 1857 and 1858. The section on the action of ice is so altered and enlarged as to be practically new; for when the first edition of the "Principles" was published comparatively little was known of the effects of land-ice, and the art of following the trail of vanished glaciers had yet to be learnt. But, with this exception, the part of the book dealing with the action of the forces of Nature--heat and cold, rain, rivers, and sea--remains comparatively unaltered, as do the first five chapters, which give a sketch of the early history of the science of geology.

Without some knowledge of this history it is hardly possible to appreciate the true greatness of the "Principles," and its unique value as an influence on scientific thought at the time it appeared. This, however, to some extent may be inferred from those chapters which we have mentioned; but the perspective of half a century enables us to understand it better at the present time; for the author, of course, had to deal with contemporary work and opinion only in a very indirect way. We may dismiss briefly the crude speculations of the earliest observers--those anterior to the Christian era--of which the author gives a summary in the second chapter of the "Principles"; for at that early date few persons had made any effort to arrange the facts of Nature in a connected system. These were too scanty and too disconnected for any such effort to be successful. The general result cannot be better summed up than in Lyell's own words:--

"Although no particular investigations had been made for the express purpose of interpreting the monuments of ancient changes, they were too obvious to be entirely disregarded; and the observation of the present course of Nature presented too many proofs of alterations continually in progress on the earth to allow philosophers to believe that Nature was in a state of rest, or that the surface had remained and would continue to remain, unaltered. But they had never compared attentively the results of the destroying and the reproductive operations of modern times with those of remote eras; nor had they ever entertained so much as a conjecture concerning the comparative antiquity of the human race, or of living species of animals and plants, with those belonging to former conditions of the organic world. They had studied the movements and positions of the heavenly bodies with laborious industry, and made some progress in investigating the animal, vegetable, and mineral kingdoms; but the ancient history of the globe was to them a sealed book, and though written in characters of the most striking and imposing kind, they were unconscious even of its existence."[56]

The above remarks hold good for the centuries immediately succeeding the Christian era; and the influence of the new faith, when it ceased to be persecuted and became a power in the state, was adverse on the whole to progress in physical or natural science. With the decline of the Roman empire a great darkness fell upon the civilised world; art, science, literature withered before the hot breath of war and rapine, as the northern barbarians swept down upon their enfeebled master on their errand of destruction. It was well nigh eight centuries from the Christian era before the spirit of scientific enquiry and the love of literature began to awaken from their long torpor; and it was then among people of an Eastern race and an alien creed. The caliphs of Bagdad encouraged learning, and the students of the East became familiar by means of translations with the thoughts and questionings of ancient Greece and Rome. The efforts of their earliest investigators have not been preserved, but in treatises of the tenth century--written by one Avicenna, a court physician, the "Formation and Classification of Minerals" is discussed, as well as the "Cause of Mountains." In the latter attention is called to the effect of earthquakes, and to the excavatory action of streams. In the same century also, "Omar the Learned" wrote a book on "the retreat of the sea," in which he proved by reference to ancient charts and by other less direct arguments that changes of importance had occurred in the form of the coast of Asia. But even among the followers of Mohammed theology declared itself hostile to science; the Moslem doctors of divinity deemed the pages of the Koran, not the book of Nature, man's proper sphere of research, and considered these difficulties ought to be settled by a quotation from the one rather than by facts from the other. So progress in science was impeded, and recantations at the bidding of ecclesiastics are not restricted to the annals of Christian races. But men seem to have gone on speculating, and Mohammed Kazwini, in a striking allegory which is quoted by Lyell, tells his readers how (to use the words of Tennyson)[57]:--

"There rolls the deep where grew the tree. O Earth, what changes thou hast seen! There, where the long street roars, hath been The stillness of the central sea."

In Europe geological phenomena do not appear to have attracted serious attention till the sixteenth century, when the significance of fossils became the subject of an animated controversy in Italy. At that epoch this country held the front rank in learning and the arts, and an inquiry of that nature arose almost as a matter of course, because the marls, sands, and soft limestones of its lower districts teem in many places with shells and other marine organisms in a singular state of perfection and preservation. It is interesting to remark, that among the foremost in appealing to inductive processes for the explanation of these enigmas was that extraordinary and almost universal genius, Leonardo da Vinci. He ridiculed the current idea that these shells were formed "by the influence of the stars," calling attention to the mud by which they were filled, and the gravel beds among which they were intercalated, as proof that they had once lain upon the bed of the sea at no great distance from the coast. His induction rested on the evidence of sections which had been exposed during his construction of certain navigable canals in the north of Italy. Shortly afterward, the conclusions of Leonardo were amplified, and strengthened on similar grounds by Frascatoro. He, however, not only demonstrated the absurdity of explaining these organic structures by the "plastic force of Nature"--a favourite refuge for the intellectually destitute of that and even a later age, but he also showed that they could not even be relics of the Noachian deluge. "That inundation, he observed, was too transient; it had consisted principally of fluviatile waters; and if it had transported shells to great distances, must have strewed them over the surface, not buried them at vast depths in the interior of mountains." As Lyell truly remarks, "His clear exposition of the evidence would have terminated the discussion for ever, if the passions of man had not been enlisted in the dispute; and even though doubts should for a time have remained in some minds, they would speedily have been removed by the fresh information obtained almost immediately afterwards, respecting the structure of fossil remains, and of their living analogues." But the difficulties raised by theologians, and the general preference for deductive over inductive reasoning, greatly impeded progress. It was not till the methods of the schoolmen yielded place to those of the natural philosophers that the tide of battle began to turn, and science to possess the domains from which she had been unjustly excluded. For about a century the weary war went on; the philosophers of Italy leading the van, those of England, it must be admitted, for long lagging behind them, before the spectre of "plastic force" was finally dismissed to the limbo of exploded hypotheses in England. For instance, it was seriously maintained by the well-known writer on county history, Dr. Plot, in the last quarter of the seventeenth century, though its absurdity had been demonstrated by his Italian contemporaries; as by Scilla, in his treatise on the fossils of Calabria, and by Steno, in that on "Gems, crystals, and organic petrifactions enclosed in solid rocks." The latter had proved by dissecting a shark recently captured in the Mediterranean, that its teeth and bones corresponded exactly with similar objects from a fossil in Tuscany, and that the shells discovered in sundry Italian strata were identical with living species, except for the loss of their animal gluten and some slight mineral change. Moreover, he had distinguished, by means of their organic remains, between deposits of a marine and of a fluviatile character.

But now, as the "plastic force" dogma lost its hold on the minds of men, its place was taken by that which regarded all fossils as the relics of an universal deluge.

"The theologians who now entered the field in Italy, Germany, France, and England, were innumerable; and henceforward, they who refused to subscribe to the position that all marine organic remains were proofs of the Mosaic deluge, were exposed to the imputation of disbelieving the whole of the sacred writings. Scarce any step had been made in approximating to sound theories since the time of Frascatoro, more than a hundred years having been lost in writing down the dogma that organised fossils were mere sports of Nature. An additional period of a century and a half was now destined to be consumed in exploding the hypothesis that organised fossils had all been buried in the solid strata by Noah's flood."[58]

Into the varying fortunes of this second struggle it is needless to enter at any length. It was the old conflict between theology and science in a yet more acute form; the old warfare between deductive and inductive reasoning; between dogmatic ignorance and an honest search for truth. Protestants and Romanists alike seemed to claim the gift of infallibility, with the right to decide ex cathedr?on questions of which they were profoundly ignorant, and to pronounce sentence in causes where they could not even appreciate the evidence. Ecclesiastics scolded; well-meaning though incompetent laymen echoed their cry; the more timorous among scientific men wasted their time in devising elaborate but futile schemes of accommodation between the discoveries of geology and the supposed revelations of the Scriptures; the stronger laboured on patiently, gathering evidence, strengthening their arguments and dissecting the fallacies by which they were assailed, until the popular prejudice should be allayed and men be calm enough to listen to the voice of truth. It was a long and weary struggle, which is now nearly, though not quite, ended; for there are still a few who mistake for an impregnable rock that which is merely the shifting-sand of popular opinion, and cannot realise that the province of revelation is in the spiritual rather than in the material, in the moral rather than in the scientific order. The outbursts of denunciation aroused by the assertion of the antiquity of man and the publication of the "Origin of Species," which many still in the full vigour of their powers can well remember, were but a recrudescence of the same spirit, a reappearance of an old foe with a new face.

But when Lyell was young and the idea of the "Principles" began to germinate in his mind, popular prejudice against the free exercise of inquiry in geology was still strong; this diluvial hypothesis still hampered, if it did not fully satisfy, the majority of scientific workers. Here and there, it is true, some isolated pioneer demonstrated the impossibility of referring the fossil contents of the earth's crust to a single deluge, or protested against the singular mixture of actual observation, patristic quotation, and deductive reasoning which commonly passed current for geological science. Chief and earliest among these men, Vallisneri, also an Italian, about a century before Lyell's birth, was clearsighted enough to see "how much the interests of religion as well as those of sound philosophy had suffered by perpetually mixing up the sacred writings with questions in physical science"; indeed, he was so far advanced as to attempt a general sketch of the marine deposits of Italy, with their organic remains, and to arrive at the conclusion that the ocean formerly had extended over the whole earth and after remaining there for a long time had gradually subsided. This conclusion, though inadequate as an expression of the truth, was much more philosophical than that of an universal and comparatively recent deluge. Moro and Generelli, in the same country, followed the lead of Vallisneri, in seeking for hypotheses which were consistent with the facts of Nature, Generelli even arriving at conclusions which, in effect, were those adopted by Lyell, and have been thus translated by him:

"Is it possible that this waste should have continued for six thousand and perhaps a greater number of years, and that the mountains should remain so great unless their ruins have been repaired? Is it credible that the Author of Nature should have founded the world upon such laws as that the dry land should be for ever growing smaller, and at last become wholly submerged beneath the waters? Is it credible that, amid so many created things, the mountains alone should daily diminish in number and bulk, without there being any repair of their losses? This would be contrary to that order of Providence which is seen to reign in all other things in the universe. Wherefore I deem it just to conclude that the same cause which, in the beginning of time, raised mountains from the abyss, has down to the present day continued to produce others, in order to restore from time to time, the losses of all such as sink down in different places, or are rent asunder, or in other ways suffer disintegration. If this be admitted, we can easily understand why there should now be found upon many mountains so great a number of crustacea and other marine animals."

This attempt at a system of rational geology was a great advance in the right direction, though many gaps still remained to be filled up and some errors to be corrected; such for instance as the idea adopted by Generelli from Moro, and maintained in other parts of his work, that all the stratified rocks are derived from volcanic ejections. Nevertheless, geology, by the middle of the eighteenth century, had evidently begun to pass gradually, though very slowly, from the stage of crude and fanciful hypotheses to that of an inductive science. But even then the observers had only succeeded in setting foot on the lower slopes of a peak, the summit of which will not be reached, if indeed it ever be, for many a long year to come. During the next half of the century progress was made, now in this direction, now in that; slowly truths were established, slowly errors dispelled; and as the close of that century approached, the foundations of modern geology began to be securely laid. A great impulse was given to the work, though to some extent the apparent help proved to be a real hindrance, by that famous teacher, Werner of Freiberg, in Saxony. His influence was highly beneficial, because he insisted not only on a careful study of the mineral character of rocks, but also on attending to their grouping, geographical distribution, and general relations. It was hurtful almost to as great a degree, because he maintained, and succeeded by his enthusiasm and eloquence in impressing on his disciples, most erroneous notions as to the origin of basalts and those other igneous rocks which were formerly comprehended under the name "trap." Such rocks he stoutly asserted to be chemical precipitates from water, and, besides this, he held views in general strongly opposed to anything like the action of uniform causes in the earth's history. In short, the Saxon Professor was in many respects the exact antithesis of Lyell, and the points of essential contrast cannot be better indicated than in the words of the latter.[59]

"If it be true that delivery be the first, second, and third requisite in a popular orator, it is no less certain that to travel is of first, second, and third importance to those who desire to originate just and comprehensive views concerning the structure of our globe. Now Werner had not travelled to distant countries; he had merely explored a small portion of Germany, and conceived, and persuaded others to believe, that the whole surface of our planet and all the mountain-chains in the world were made after the model of his own province. It became a ruling object of ambition in the minds of his pupils to confirm the generalisations of their great master, and to discover in the most distant parts of the globe his 'universal formations,' which he supposed had been each in succession simultaneously precipitated over the whole earth from a common menstruum or chaotic fluid."

These wild generalisations, as Lyell points out, had not even the merit of being really in accordance with the evidence afforded by some parts of Saxony itself. Werner, in fact, was a conspicuous example of a tendency, which perhaps even now is not quite extinct, to work too much beneath a roof and too little in the open air; to found great generalisations on the minute results of research in a laboratory, without subjecting them to actual tests by the study of rocks in the field.

This error on Werner's part was the less excusable, because, even before he began to lecture, the true nature of basalts and traps generally had been recognised by several observers of different nationalities. In the Hebrides and in Iceland, in the Vicentin and in Auvergne, even in Hesse and in the Rheingau, proof after proof had been cited, and the evidence in favour of the "igneous" origin of these rocks had become irresistible, as one might suppose, within some half dozen years of Werner's appointment as professor at Freiberg. Faujas, in 1779, published a description of the volcanoes of the Vivarais and Velay, in which he showed how the streams of basalt had poured out from craters which still remain in a perfect state. Desmarest also pointed out that in Auvergne "first came the most recent volcanoes, which had their craters still entire and their streams of lava conforming to the level of the present river courses. He then showed that there were others of an intermediate epoch, whose craters were nearly effaced, and whose lavas were less intimately connected with the present valleys; and lastly, that there were volcanic rocks still more ancient without any discernible craters or scori? and bearing the closest analogy to rocks in other parts of Europe, the igneous origin of which was denied by the school of Freiberg." Desmarest even constructed and published a geological map of Auvergne, of which Lyell speaks in terms of high commendation. "They alone who have carefully studied Auvergne, and traced the different lava streams from their craters to their termination--the various isolated basaltic cappings--the relation of some lavas to the present valleys--the absence of such relations in others--can appreciate the extraordinary fidelity of this elaborate work."[60]

But before the close of the eighteenth century, two champions had already stepped into the arena to withstand the Wernerian hypothesis, which, like a swelling tide, was spreading over Europe, and threatening to sweep away everything before it. These were James Hutton and William Smith; the one born north, the other south of the Tweed. From the name of the former that of his friend and expositor, John Playfair, must never be separated. They were the Socrates and the Plato of that school of thought from which modern geology has been developed.[61] To quote the eloquent words of Sir Archibald Geikie[62]:--

"On looking back to the beginning of this century we see the geologists of Britain divided into two hostile camps, which waged against each other a keen and even an embittered warfare. On the one hand were the followers of Hutton of Edinburgh, called from him the Vulcanists, or Plutonists; on the other, the disciples of Werner ... who went by the name of Wernerians, or Neptunists.... The Huttonians, who adhered to the principles laid down by their great founder, maintained, as their fundamental doctrine, that the past history of our planet is to be explained by what we can learn of the economy of Nature at the present time. Unlike the cosmogonists, they did not trouble themselves with what was the first condition of the earth, nor try to trace every subsequent phase of its history. They held that the geological record does not go back to the beginning, and that therefore any attempt to trace that beginning from geological evidence was vain. Most strongly, too, did they protest against the introduction of causes which could not be shown to be a part of the present economy. They never wearied of insisting that to the everyday workings of air, earth, and sea, must be our appeal for an explanation of the older revolutions of the globe. The fall of rain, the flow of rivers, the slowly crumbling decay of mountain, valley, and shore, were one by one summoned as witnesses to bear testimony to the manner in which the most stupendous geological changes are slowly and silently brought about. The waste of the land, which they traced everywhere, was found to give birth to soil--renovation of the surface thus springing Phoenix-like out of its decay. In the descent of water from the clouds to the mountains, and from the mountains to the sea, they recognised the power by which valleys are carved out of the land, and by which also the materials worn from the land are carried out to the sea, there to be gathered into solid stone--the framework of new continents. In the rocks of the hills and valleys they recognised abundantly the traces of old sea-bottoms. They stoutly maintained that these old sea-bottoms had been raised up into dry land from time to time by the powerful action of the same internal heat to which volcanoes owe their birth, and they pointed to the way in which granite and other crystalline rocks occur as convincing evidence of the extent to which the solid earth has been altered and upheaved by the action of these subterranean fires."

Such were the leading principles of the "Huttonian theory," though perhaps they are stated here in a slightly more developed form than when it was first presented by its illustrious author. But it was defective in one important respect, on a side from which it might have obtained the strongest support, and have liberated itself from the bondage of deluges; in other words, of convulsive action, by which it was still fettered, for "it took no account of the fossil remains of plants and animals. Hence it ignored the long succession of life upon the earth which those remains have since made known, as well as the evidence thereby obtainable as to the nature and order of physical changes, such as alternations of sea and land, revolutions of climate, and suchlike."

This defect was supplied by William Smith. He had learnt, by patient labour among the stratified rocks of England, to recognise their fossils, had ascertained that certain assemblages of the latter characterised each group of strata, and by this means had traced such groups through the country, and had placed them in order of superposition. So early as 1790, he published a "Tabular View of the British Strata," and from that time was engaged at every spare moment in constructing a geological map of England, all the while freely communicating the results of his researches to his brethren of the hammer. "The execution of his map was completed in 1815, and it remains a lasting monument of original talent and extraordinary perseverance; for he had explored the whole country on foot without the guidance of previous observers, or the aid of fellow labourers, and had succeeded in throwing into natural divisions the whole complicated series of British rocks."[63]

A most important step in view of future progress, at any rate in our own country, was taken by the foundation of the Geological Society of London in 1807, the members of which devoted themselves at first rather to the collection of facts than to the construction of theories, while in France the labours of Brongniart and Cuvier in comparative osteology, and of Lamarck in recent and fossil shells, smoothed the way toward the downfall of catastrophic geology. Those men, with their disciples, "raised these departments of study to a rank of which they had never before been deemed susceptible. Their investigations had eventually a powerful effect in dispelling the illusion which had long prevailed concerning the absence of analogy between the ancient and modern state of our planet. A close comparison of the recent and fossil species, and the inferences drawn in regard to their habits, accustomed the geologist to contemplate the earth as having been at successive periods the dwelling-place of animals and plants of different races--some terrestrial, and others aquatic; some fitted to live in seas, others in the waters of lakes and rivers. By the consideration of these topics the mind was slowly and insensibly withdrawn from imaginary pictures of catastrophes and chaotic confusion, such as haunted the imagination of the early cosmogonists. Numerous proofs were discovered of the tranquil deposition of sedimentary matter, and the slow development of organic life."[64]

Such was the earlier history of Geology; such were the influences which had moulded its ideas till within a few years of the date when Lyell began to make it a subject of serious study. At that time, namely about the year 1820, the Geological Society of London had become the centre and meeting-point of a band of earnest and enthusiastic workers, whose names will always hold an honoured place in the annals of the Science. Among the older members--most of whom, however, were still in the prime of life, were such men as Buckland, Conybeare, Fitton, Greenough, Horner, MacCulloch, Warburton and Wollaston; among the younger, De la Beche and Scrope, Sedgwick and Whewell. Murchison, though a few years Lyell's senior, was by almost as many his junior as a geologist, for he did not join the Society till the end of 1824, and was actually admitted on the evening when Lyell, then one of its honorary secretaries, read his first paper--on the marl-lake at Kinnordy. Such men also as Babbage, Herschel, Warburton, Sir Philip Egerton, the Earl of Enniskillen (then Viscount Cole), must not be forgotten, who were either less frequent visitors or more directly devoted to other studies. At this time geology was passing into a phase which endured for some forty years--the exaltation of the paleontological, the depreciation of the mineralogical side. If it be true, as it has been more than once remarked, that the father of the geologist was a mineralogist, it is no less true that his mother was a paleontologist; but at this particular epoch the paternal influence obviously declined, while that of the mother became inordinately strong. Wollaston and MacCulloch, indeed, were geologists of the old school; excellent mineralogists and petrologists (to use the more modern term) as accurate as it was possible to be with the appliances at their disposal, but among the younger men De la Beche, accompanied to a certain extent by Scrope and Sedgwick, was almost alone in following their lead. But although paleontology and stratigraphical geology as its associate were clearly making progress, the school of thought, of which Lyell became the champion, counted at this time but few adherents, for the older geologists were almost to a man "catastrophists." A few, like MacCulloch, undervalued paleontological research, and thus were doubly prejudiced against the uniformitarian views. Buckland, Conybeare, Greenough, as we have already seen from incidental remarks in Lyell's letters, had put their trust in deluges, and imagined that by such an agency the earth had been prepared for a new creation of living things and a new group of geological formations. Sedgwick even was to a great extent on their side. He had speedily emerged from the waters of Wernerism, in which at first he had been for a short time immersed, but he did not escape so easily from the roaring floods of diluvialists, and the grandeur of catastrophic changes in the crust of the earth fascinated his enthusiastic, almost poetic, nature. Even so late as 1830, we find him criticising from the chair of the Geological Society the leading argument of Lyell's "Principles of Geology" in no friendly spirit, and bestowing high praise on Elie de Beaumont's theory of Parallel Mountain-chains.

A brief summary of the views advocated by this eminent French geologist may serve to indicate, perhaps better than any general statements, the influences against which Lyell had to contend at the outset of his career as a geologist. With the omission of certain parts, to which no exception would be taken, or which have no very direct bearing upon the immediate question, they are as follows[65]: (1) In the history of the earth there have been long periods of comparative repose, during which the sedimentary strata have been continuously deposited, and short periods of paroxysmal violence, during which that continuity has been interrupted. (2) At each of these periods of violence or revolution in the state of the earth's surface, a great number of mountain-chains have been formed suddenly, and these chains, if contemporaneous, are parallel; but if not so, generally differ in direction. (3) Each revolution or great convulsion has coincided with the date of another geological phenomenon, namely, the passage from one independent sedimentary formation to another, characterised by a considerable difference in "organic types." (4) There has been a recurrence of these paroxysmal movements from the remotest geological periods; and they may still be produced.

Thus the force of authority, which has to be reckoned with in geology, if not in other branches of science, was in the main adverse to Lyell, who could count on but few to join him in his attack on catastrophism. One indeed there was, a host in himself, who, though his contemporary in years, had devoted himself wholly to geology at a slightly earlier date and had already become convinced, by his field-work in Italy and France, of the efficacy of existing forces to work mighty changes, if time were given, in the configuration of the earth's surface. This was George Poulett Scrope, a man of broad culture, great talents, and singular independence of thought, who had convinced himself of the errors of the Wernerian theory by his studies in Italy in the years 1817-19, and had thoroughly explored the volcanic district of Auvergne in 1821. His work on the Phenomena of Volcanoes, published in 1823, and that on the Geology of Central France, published in 1826, had given the coup de grace to Werner's hypothesis and had made the first breach in the fortress of the catastrophists.

For a complete solution of the problem to which Lyell had addressed himself, two methods of investigation were necessary. It must be demonstrated that in tracing back the life history of the earth from the present age to a comparatively remote past no breach of continuity could be detected, and that the forces which were still engaged in sculpturing and modifying this earth's surface were adequate, given time enough, to produce all those changes to which the catastrophist appealed as proofs of his hypotheses. To establish the one conclusion, it was necessary to make a careful study of the Tertiary formations, which were still in a condition of comparative confusion; to arrange them in an order no less clear and definite than that of the Secondary systems; and to show, by working downward from the present fauna, not only that many living species had been long in existence, but also that these had appeared gradually, not simultaneously, and had in like manner replaced forms which had one after another vanished--to prove, in short, "that past and present are bound together by an unsevered cord of life, whose interlacing strands carry us back in orderly change from age to age." To establish the other conclusion it was necessary to show that, even in historical times, considerable changes had occurred in the outlines of coasts, and that heat and cold, the sea, or rain and rivers--especially the last--had been agents of the utmost importance in the sculpture of cliffs, valleys, and hills. For both these purposes careful study, not only in Britain, but also still more in other regions, was absolutely necessary, and it was with them in view that Lyell undertook his journeys, from the time when his geological ideas began to assume a definite shape until the last volume of the "Principles" was published. By that date, as has been stated in the preceding chapters, he had made himself familiar in the course of his geological education with many parts of Britain, had laboriously investigated the more important collections and museums of France and Italy, and had carefully studied in the field the principal Tertiary deposits not only in these countries but also in Sicily and in parts of Switzerland and Germany. To obtain evidence bearing on the physical aspect of the question on a scale grander than was afforded by the undulating lowlands, or worn-down highland regions of Britain and the neighbouring parts of Europe, he had rambled among the Alps and Pyrenees, examining their peaks and precipices, their snowfields, glaciers, lakes, and torrents, and watching the processes of destruction, transportation, and deposition of which crag, stream, and plain afford a never-ending object-lesson. In order to study volcanoes still in activity, he had climbed Vesuvius and Etna; in order to scrutinise more minutely the structure of cones, craters, and lava streams, he had visited Auvergne, Catalonia, and the Eifel; while in all his goings and comings through scenes where Nature worked more unobtrusively, he had watched her never-ending toil, as she destroyed with the one hand and built with the other. He was thus able to write with the authority of one who has seen, not of one who merely quotes; of one who knew, not of one who had learnt by rote. The "Principles of Geology," though of course it had to rely not seldom on the work of others, bore the stamp of the author's experience, and was redolent, not of the dust of libraries, but of the sweetness of the open air. That fact added no little force to its cautious and clear inductive reasoning; that fact did much to disarm opposition, and to open the way to victory.

FOOTNOTES:

[55] Strictly speaking, fifteen out of the last sixteen chapters, for the final one (dealing with coral reefs) is substantially a reprint.

[56] "Principles of Geology," vol. i. p. 26 (eleventh edition).

[57] In Memoriam, cxxiii.

[58] "Principles of Geology," chap. iii. p. 37.

[59] "Principles of Geology," chap. iv.

[60] "Principles of Geology," chap. iv.

[61] Hutton's "Theory of the Earth" was first published in 1788, and in an enlarged form in 1795. Playfair's "Illustrations of the Huttonian Theory" appeared in the spring of 1802.

[62] Geikie's "Life of Murchison," chap. vii.

[63] "Principles of Geology," chap. iv.

[64] "Principles of Geology," chap. iv.

[65] Abridged from Lyell's summary: "Principles of Geology," chap. vii.

CHAPTER VI.

EIGHT YEARS OF QUIET PROGRESS.

Both courses of lectures ended[66] and the third volume of the "Principles" successfully launched, Mr. and Mrs. Lyell left London in June, 1833, for another Continental tour. During their first halt, at Paris, she was duly introduced to the famous quarries of Montmartre, and had an opportunity of "collecting a fossil shell or two for the first time." Thence they made their way to Bonn, which she had left as a bride the previous summer, and, after another short halt, proceeded up the gorge of the Rhine to Bingen, visiting on the way the ironworks at Sayn, and examining the stratified volcanic deposits on the plain between the river and that town. The Tertiary basin at Mayence was next visited, and from it they went leisurely to Heidelberg. From the picturesque old town by the Neckar they struck off to Stuttgart and to Pappenheim, examining one or two collections at the former place, and the quarries of Solenhofen, near the latter. These were already noted for the abundant and well-preserved fossils obtained in the quarries worked for the well-known "lithographic stone," though the famous Archaeopteryx had yet to be found; that strange creature, feathered and like a bird, but with teeth in its beak and a tail like a reptile, which has supplied such an important link in the chain of evidence in favour of progressive development. Thence they travelled to Nurnberg and Bayreuth, visiting on their way the noted caves at Muggendorf, and returned to Bonn by way of Bamberg, Waltzburg, Aschaffenberg, and Frankfurt. In this journey, few localities of special interest were investigated, but, as Lyell's letters show, no opportunity was lost of discussing important questions with local geologists, or of examining sections in the field. But on the way back to England through Belgium a halt was made at Liege, to inspect Dr. Schmerling's grand collection of cave-remains. It is evident, though but a short notice of it has been preserved, that this visit kindled an enthusiasm which was to produce important results in later years. Lyell writes (to Mantell, after his return to England):--

"I saw at Liege the collection of Dr. Schmerling, who in three years has, by his own exertion and the incessant labours of a clever amateur servant, cleared out some twenty caves untouched by any previous searcher, and has filled a truly splendid museum. He numbers already thrice the number of fossil cavern mammalia known when Buckland wrote his 'Idola Specus'; and such is the prodigious number of the individuals of some species--the bears, for example, of which he has five species, one large, one new--that several entire skeletons will be constructed. Oh, that the Lewes chalk had been cavernous! And he has these, and a number of yet unexplored and shortly to be investigated holes, all to himself: but envy him not--you cannot imagine what he feels at being far from a metropolis which can afford him sympathy; and having not one congenial soul at Liege, and none who take any interest in his discoveries save the priests--and what kind they take you may guess, more especially as he has found human remains in breccia, embedded with the extinct species, under circumstances far more difficult to get over than any I have previously heard of. The three coats or layers of stalagmite cited by me at Choquier are quite true."[67]

Very probably among these human relics was one which was destined to become famous--the skull found in the cave at Engis--for this was described by Dr. Schmerling in his "Recherches sur les ossements fossiles douverts dans les cavernes de la Province de Liege," a book published in 1833. It was found at a depth of nearly five feet, hidden under an osseous breccia, composed of the remains of small animals, and containing one rhinoceros tusk with several teeth of horses and of ruminants. The earth in which it was lying did not show the slightest trace of disturbance, and teeth of rhinoceros, horse, hydra, and bear surrounded it on all sides.[68] This relic proved--and since then numbers of similar cases have been discovered--that if the man of Engis were an antediluvian, and his corpse had been washed into the cave together with the drowned bodies of rhinoceros, and other animals,[69] that event, at any rate, must have corresponded with a great change in the habits of the larger mammalia, for they had been unable to return to haunts which once had been congenial. In other words, the foundation was being laid, now in 1833, for the next great advance in geological science, the contemporaneity of man and several extinct species of mammals, indicating, of course, the antiquity of the human race. To this point, however, public attention was not directed for nearly twenty years. Then various causes, especially an examination into the evidence discovered in the neighbourhood of Abbeville and Amiens by M. Boucher de Perthes, brought the question to the front. But though the controversy was sharp and bitter for a time, it was speedily over, and the question which is still agitated--though mildly and in a sense wholly scientific--is whether man appeared in this part of Europe and in corresponding regions of North America, before, during, or after the glacial epoch?

But the Engis skull is a relic exceptionally interesting. Though the handiwork of primeval man is common enough--rudely chipped instruments or weapons of flint or other stone, worked portions of bones and antlers, and such like--yet his bones are far less common than those of other mammals, and, most of all, skulls are rare. Professor Huxley, in his work from which we have already quoted, states that Dr. Schmerling found a bone implement in the Engis cave, and worked flints in all the ossiferous Belgian caves, yet this was the only skull in anything like a perfect condition, though another cavern furnished two fragments of parietal bones. Yet from the latter numerous bones of the extremities were obtained, and these had belonged to three individuals. What inferences, then, can be drawn from this skull as to the intellectual rank of primeval man? This question was discussed by its discoverer, and the evidence has been also considered by Professor Huxley. The former thus expressed his opinion, "that this cranium has belonged to a person of limited intellectual faculties, and we conclude thence that it belonged to a man of a low degree of civilisation; a deduction which is borne out by contrasting the capacity of the frontal with that of the occipital region." Professor Huxley sums up a careful discussion of the evidence, in which he calls special attention to points where it happens to be defective, by stating that the specimen agrees in certain respects with Australian skulls, in others with some European, but that he can find in the remains no character which, if it were a recent skull, would give any trustworthy clue to the race to which it might appertain. "Assuredly there is no mark of degradation about any part of its structure. It is, in fact, a fair average human skull, which might have belonged to a philosopher, or might have contained the thoughtless brains of a savage."[70]

The winter of 1833 and the spring of the following year were spent in London. It was evidently a busy, though uneventful, time: a new edition of the "Principles" was being prepared and printed, a paper read to the Geological Society on a freshwater formation at Cerdagne in the Pyrenees, and information collected for a summer's journey. This was to be in a new direction--to Scandinavia--with the more especial intent of studying the evidence on which it has been asserted that the shores of the Baltic had changed their level within recent times. But on this occasion Mrs. Lyell remained at home, as the travelling might occasionally have been too rough for her so we find, in a journal written for her perusal, a full sketch of a tour which proved, as he had anticipated, to be fruitful in scientific results. His first halt was at Hamburg, where, on his arrival, with characteristic energy he dashed off at once in a carriage to examine a section below Altona which he had marked down on his voyage up the Elbe. This is his brief summary: "Cliffs sixty or seventy feet high. Filled three pages of note-book. Saw the source of the great Holstein granite blocks. Gathered shells thrown ashore by the Elbe." From Hamburg he drove to Leck, along one of the worst of roads. The primary cause of its badness was geological--a loose sand interspersed with granite boulders; the secondary, the royal revenues; for these largely depended on the tolls paid by vessels on entering the sound, and if a good road had connected the two towns much merchandise would have gone overland, to the king's loss. At Leck Lyell for the first time stood upon the shore of the Baltic, and utilised the half-hour before his steamer started for Copenhagen by hunting for shells. As a reward, he found a well-known freshwater genus (Paludina) among common marine forms.[71]

From Copenhagen a rapid journey in Seeland and introduced him to a number of interesting Sections of the drift, accounts of which were afterwards worked into his books, and showed him at Faxoe and elsewhere limestones overlying the upper chalk, like those at Maestricht in Holland, and at Meudon near Paris. All these limestones possess an exceptional interest, for they contain a mixture of Secondary with Tertiary fossils, and thus help to fill up the wide gap between these two great divisions in Britain and the adjacent parts of Europe. On his return to Copenhagen Lyell was very kindly received by the Crown Prince, who was an ardent naturalist, and allowed him to examine a fine collection of minerals and fossils accumulated by himself.

After crossing the Sound to Malm? Lyell spent about a fortnight in driving along an inland route through the southern part of Sweden to Norrking, while a halt at Lund afforded the opportunity of pleasant talks with the professors of the University, and of seeing some formations of which hitherto he had not had much experience. The terms in which he refers to these indirectly proves what strides geology has taken in the last sixty years. "We made an excursion together through a country of greywacke with orthoceratite limestone and schist,[72] containing a curious zoophyte called graptolite in great abundance, and a few shells." On the journey also he found much to interest a geologist--boulders almost everywhere, some of huge size, lying on the surface or scattered in the sand in one place an outcrop of Cretaceous greensand, full of belemnites, which were popularly regarded as "witches' candles." Then over a picturesque granite region--"a country of rock, fir-wood, and peasants"--till he arrived at Norrking, and made his way in a steamer down one fjord and up another until he came into the Malar Lake. These last stages introduced him to a kind of scenery of which Scandinavia affords such striking and innumerable examples--the margin of a submerged mountain land. "We entered," he says, "a passage between an endless string of islets and the mainland, the water here smooth as a millpond. We passed swiftly on in deep water close to the rocks, on the barest of which are a few firs in the clefts. These are evidently the summits of submarine mountains." At Stockholm he found plenty to be done. Some of the evidence, which had been brought forward to prove a rising of the land, was obviously weak. For instance, on one of his first visits to a place where the upward movement was said to be comparatively rapid, he found a fine oak-tree, perhaps a couple of centuries old, growing eight feet above high-water mark, and thus indicating either that oak-trees had recently changed their habits or that the change of level had been slow. "In dealing with this question it is necessary," he writes, "to cross-examine both nature and man. The testimony of the former is strong; of the latter, I must say, so weak and contradictory that I require to know the men and find how they got their views." A valuable precaution this, which might be remembered with advantage in days when stay-at-home geologists are far too numerous. If this were done, the paper currency of the science would be considerably reduced in quantity, and there would be a closer correspondence between its real and its nominal value. A little scepticism was certainly justifiable, for one would-be savant stood him out "that a bed of Cardium edule (the common cockle) 100 feet high proves that the fresh water of Lake Malar was once that much higher." Lyell adds nothing to this remark, but his silence is eloquent.

This expedition, however--to Sertelje--gave results yet more striking than marine shells 100 feet above the present level of the Baltic. "What think you," he writes, "of ships in the same formation, nay, a house? It is as true as the Temple of Serapis.[73] I do not mean that I discovered all this, but I shall be the first to give a geological account of it. I am in high spirits at the prize." Upsala also, to which he next moved, increased his stores of knowledge and of fossils. "I went to the hill, a hundred feet high, on which the tower stands, to examine marine shells. All of Baltic species. You remember that in the half-hour between the two steamboats at Leck, or rather Travemunde, I collected shells by the quay. Not one fossil have I found newer than the chalk in Sweden, that was not in the number of those found living in that half-hour." More localities for shells were visited, erratics were examined, and pilots were questioned closely "about the agency of ice, in which they believe." With their opinion Lyell inclined to agree; at any rate, he was convinced that his observations would "quite overset the theory," and, as he expected, "bring in ice carriage as the cause." On the coast further north at Oregrund and Gefle, bench-marks had been cut some years previously in order to apply a more exact test to the question of the change in levels. These he visited, and the former seemed to prove "as Galileo said in a different sense, that 'the earth moves.'" The marks near Gefle afforded similar testimony, so that he felt now that the main object of his journey was accomplished, and inserted this pregnant note in his journal:--"I feel now what I was very sensible of when correcting my last edition,[74] that I was not justified in writing any more until I had done all in my power to ascertain the truth in regard to the 'great northern phenomenon,' as the gradual rise of part of Sweden has been very naturally called. You will see by-and-by how important a point it was, and how materially it will modify my mode of treating the science, and how much it will advance the theory of the agency of existing causes as a key to explain geological phenomena."[75]

But the work at sea-marks was not yet quite ended, and there was besides another classic spot to be visited--Uddevalla, between Lake Werner and the western coast. Here are deposits in which sea-shells are abundant at a height of about two hundred feet above the sea. Nothing but a submergence can account for their presence, for polyzoa and barnacles are found attached to the solid rock. Some of the latter, adhering to the gneiss, were collected by Lyell on this occasion.[76]

Fossil shells (of existing species) were so numerous that, he says, the deposit was worked for making lime, and he compares it with a well-known bed in the Tertiaries of the Paris Basin. The shells, however, at Uddevalla, as he points out, are not of that brackish-water character peculiar to the Baltic, but such as now live in the Northern Ocean.[77] On reaching the coast he made an expedition by boat, and saw the bench-mark at Gullholmen, and rocks which had emerged from the sea within the memory of people still living. Here, by way of completing his work, he "hired the services of a smith to make a mark at the water's edge:--

C. 18. L. ------ 18. 7. 34." -----------

So he brought his journey in Scandinavia to a close, and by the end of July had reached Kinnordy, where Mrs. Lyell awaited his coming. Then he set to work to prepare a brief sketch of his investigations for the approaching meeting of the British Association in Edinburgh, and a more elaborate paper, to be communicated to the Royal Society in London, in which he set forth the reasons which had convinced him that in Sweden, "both on the Baltic and ocean side, part of that country is really undergoing a gradual and insensibly slow rise." It affects an area measuring about one thousand miles north and south, and is believed to reach a maximum at the North Cape. There it is said, but the statement needs verification, to amount to five feet in a century; at Gefle, ninety miles north of Stockholm, it cannot be more than two or three feet in the same time; while at Stockholm itself it can hardly exceed six inches. Further south, in Scania proper, as at Malm? Skan, Trelleborg, and Ystad, the movement is distinctly in an opposite direction.[78]

This paper was afterwards accepted by the Royal Society as the Bakerian lecture for the year. But the preparation of this was not Lyell's only occupation. In October he had begun fossil ichthyology, was attending lectures in chemistry, and "had made some progress," as he writes to Mantell, "in a single volume which two years ago I promised Murray, a purely elementary work for beginners in geology, and which I find more agreeable work than I had expected." So his hands were pretty full. A pleasant surprise came in the closing months of the year, namely the award of one of the Royal Medals by that Society in acknowledgment of the merits of his "Principles of Geology."

In the earlier part of 1835 Lyell accepted the presidency of the Geological Society, an office which, it will be remembered, he had virtually refused a couple of years before, when he was busy with his great book. With this exception, nothing worthy of record appears to have happened in the first six months of the year, but in July Mrs. Lyell and he left England for a journey to France, Germany, and Switzerland. By that date, as he mentions in a letter to a friend, 1,750 copies of the last edition of the "Principles" had been sold, a demand that puts him in good heart as to the future of the book, and proves that his labours on it had not been in vain. But he did not permit himself to be idle. As a letter written to Sedgwick from Paris shows, he was still working away at the classification of the Tertiary deposits; for in this letter he discusses the relation of the coralline and the red, or shelly Crag of Suffolk. Mr. Charlesworth, subsequently well known as a collector, had been obtaining a number of fossil shells from the former deposit, and the character of these suggested that it was distinctly the older of the two, as is now universally admitted. In discussing this question Lyell lays down a principle of classification the soundness of which has been proved by experience, namely, that the age of a Tertiary deposit is to be determined by the proportion of recent species and the relation of these to the forms still living in the neighbouring seas. If, for instance, the recent shells in a formation, amounting to one-half, or even as few as one-third, of the total number can be thus found, the formation will be Pliocene in age, "while the recent shells of the Miocene have a more exotic and tropical form." To this conclusion he had been led, by an examination, with the help of Deshayes, of a typical collection of Crag fossils which he had carried with him to Paris. As to other matters, the leading French geologists were still warring vigorously in defence of deluges, and none of his numerous heresies, he remarks, appears "to have excited so much honest indignation as his recent attempt to convey some of the huge Scandinavian blocks to their present destination by means of ice."

He had proved, he reminds Sedgwick, that "some of the great blocks near Upsala must have travelled to their present destination since the Baltic was a brackish water sea, so that those who maintain that there was one, and one only, rush of water, which scattered all the blocks of Sweden and the Alps, must make out this catastrophe to be, as it were, an affair of yesterday." Geology, even at that date, had advanced far enough for this admission to have landed the diluvialists in some awkward dilemmas, to say nothing of the physical difficulties which they would find in accounting for the existence of waves or currents potent enough to bowl the Pierre ?bot from the aiguilles round the Trient glacier to the slopes of the Jura, or to fling the erratics of Scandinavia broadcast over the lowlands around the Baltic. This, however, was not the only lost cause over which the French geologists were holding their shield. Lyell goes on to write, with a touch of quiet sarcasm: "As to the elevation crater business, Von Buch, de Beaumont, and Dufresnoy are to write and prove that Somma and Etna are elevation craters, and Von Buch himself has just gone to Auvergne to prove that Mont Dore is one also."

Lyell's special intention in visiting the Alps was to obtain evidence as to the relation of the metamorphic and sedimentary rocks. Geologists of the Wernerian School, with sundry others who hardly went so far as the Freiberg professor, maintained that the crystalline schists, including gneiss, had been produced, often as precipitates, in a primeval ocean, the waters of which were far too hot to allow of the existence of life. At a later time, as the temperature fell, the great masses of slightly altered slates and grits were deposited--the region of "greywacke," the transitional rocks as they were commonly called. These for the most part were unfossiliferous, at any rate in their earliest stages. To this view, of course, the Huttonian dictum, which Lyell sought to establish, was diametrically opposed, viz. that the earth showed no signs of a beginning. Now he had been informed that in the Alps certain slaty rocks contained fossils which indicated an age corresponding generally with the chalk of England, and that in other parts of that chain even crystalline schists could be found interbedded with fossiliferous strata of Secondary age. To settle the former question he intended to visit the famous quarries of Glarus, but was ultimately compelled to leave this for another year, as he took the latter point first in order of time, and the investigation of it involved more work than he had anticipated. In regard to this, the most important sections were to be found on the precipitous northern slopes of the Jungfrau and in the upper part of the Urbach-thal, a lonely glen which descends into the main valley of the Aar at Imhof, above Meyringen. In both these localities gneiss appears to overlie "fossiliferous limestone," and Lyell, after visiting them, returned satisfied that he had seen "alternations of the gneiss with limestone of the lias or something newer in the highest regions of the Alps." That undoubtedly he saw, but he did not suspect that the appearance was illusory. This was not in the least surprising; the Alps were still almost a terra incognita; the processes of "mountain making" as yet were unknown; many statements in common currency as to the passage of sedimentary into crystalline rocks were erroneous and distinctly misleading. Only by degrees was it discovered that this superposition of gneiss or crystalline schist to Secondary rock was due to folding on a scale so gigantic that the older had been doubled over upon the younger rock and the apparent order of succession was the converse of the true one. The intercalation also of the gneiss and the Jurassic limestone was a result of a similar action, but carried, if possible, to an even greater extreme, for here the hard gneiss had been thrust in wedge-like slabs between the softer masses of sedimentary rock, like a paper-knife between the leaves of a book; that is to say, the gneiss and crystalline schists in both cases were vastly more ancient than the fossiliferous limestone. It is only of late years that this startling fact has been established beyond question; and even now there are many geologists who do not appear to recognise how seriously the Huttonian dictum "there is no sign of a beginning" has been shaken by the collapse of this evidence. At the present time the question is in this position; all the attempts to prove crystalline schists to be of the same age as, or younger than, fossiliferous sedimentary rocks either have been complete failures or have proved to be very dubious, while in many cases these schists are demonstrably earlier than the oldest rocks of the district to which a date can be assigned. Hence, though possibly it may turn out that the disciples of Hutton were right, and that, as Lyell thought, a metamorphic rock may be of almost any geological age, his hypothesis not only is unproved, but also the evidence which has been brought forward in its favour has turned out after a strict scrutiny to be exceedingly dubious, if not absolutely contrary. In regard to this question we may feel a little surprise that one difficulty did not occur to Lyell's sceptical mind, namely: what could be the nature and cause of a process of metamorphism which could convert one sediment into a crystalline schist--changed practically past recognition--and leave its neighbour so far unaltered that its characteristic fossils could be readily recognised?

But though he was unable to investigate the question of Secondary or perhaps early Tertiary fossils in the "transition"-like rock of Glarus, his study of the sedimentary deposits of the Bernese Oberland, which had formed a necessary preliminary to the other inquiry, raised some difficulties in his mind as to the origin of slaty cleavage. At a meeting of the Geological Society in the month of March, Professor Sedgwick had read his classic paper[79] on this subject, in which he established the independence of cleavage and bedding. This paper laid the foundation for the discovery of the true cause of the former structure, though its author was unable, with the information then at his command, to do more than suggest an hypothesis, which afterwards proved to be incorrect. He had shown that both the strike and the dip of cleavage-planes were persistent over large areas, and that while the one might gradually change its direction and the other its angle of inclination, if they were followed far enough, yet this angle usually remained unaltered for considerable distances, and appeared to be quite unaffected by any variation in the slope of the strata. From these observations it followed that the planes of cleavage ought not to be coincident with those of bedding. Lyell, however, writes to tell Sedgwick[80]:--

"I found the cleavage or slaty structure of fine drawing slate in the great quarry of the Niesen, on the east [south] side of the Lake of Thun, quite coincided with the dip of the strata ascertained by alternate beds of greywacke.... As it is the best description of drawing slate, and as divisible almost as mica into thin plates, I cannot make out how to distinguish such a structure from any which can be called slaty, and such an attempt would, I fear, involve the subject in great confusion."

The observation was perfectly correct, and many like instances could be found in the Alps; nevertheless, Sedgwick was right in his generalisation, and the two structures are perfectly independent, though the difficulty raised by Lyell did not disappear till the true cause of slaty cleavage was recognised--viz. that it is a result of pressure. Thus, in a region like the Alps, where the strata often have been so completely folded as to be bent, so to say, back to back, the planes of cleavage, which are produced when the rocks can no longer yield to the pressure by bending, necessarily coincide with those of bedding. Still, even in these cases, if careful search be made in the vicinity, some minor flexure generally betrays the secret, and exhibits the cleavage structure cutting across that of bedding.

The next year, 1836, flowed on, like the last, quietly and uneventfully; a fifth edition of the "Principles" was passing through the press; the "Elements of Geology" was making progress, though slowly; and Lyell's duties as President of the Geological Society, which involved the delivery of an address in the month of February and the preparation of another one for the same season in the following year, occupied a good deal of his time. The summer was spent in a long visit to his parents at Kinnordy, after which he and Mrs. Lyell made some stay in the Isle of Arran before they returned to London. The latter seemingly had been rather out of health, and this may have been the reason why a longer journey was not undertaken, but she must have found the Scotch air a complete restorative, for after her return to London in the autumn Lyell writes to his father that "everyone is much struck with the improvement in Mary's health and appearance."

But one letter, of the few which have been preserved from those written in 1836, possesses a special interest, for it expresses his ideas, at this epoch, in regard to the question of the origin of species, and indicates his freedom from prejudice and the openness of his mind. It is addressed to Sir John Herschel, then engaged in his memorable investigations at the Cape of Good Hope, who had favoured him with some valuable comments and criticisms on the Principles of Geology, and in the course of these had corrected a mistake which Lyell had made in regard to a rather difficult physical question. In referring to this, the latter remarks that the clearness of the mathematical reasoning (to quote his words) "made me regret that I had not given some of the years which I devoted to Greek plays and Aristotle at Oxford, and afterwards to law and other desultory pursuits, to mathematics." Doubtless there is hardly any better foundation for geology than a course of mathematics; at the same time, classical studies did much to give Lyell his lucidity and elegance of style, and thus to ensure the success of the "Principles of Geology."

It will be best to give Lyell's own words, for the document forms an appendix or lengthy postscript. As is incidentally mentioned, it was not in his own handwriting,[81] and thus probably was drawn up with rather more than usual care.

"In regard to the origination of new species, I am very glad to find that you think it probable it may be carried on through the intervention of intermediate causes. I left this rather to be inferred, not thinking it worth while to offend a certain class of persons by embodying in words what would only be a speculation.... When I first came to the notion--which I never saw expressed elsewhere, though I have no doubt it had all been thought out before--of a succession of extinction of species, and creation of new ones, going on perpetually now, and through an indefinite period of the past, and to continue for ages to come, all in accommodation to the changes which must continue in the inanimate and habitable earth, the idea struck me as the grandest which I had ever conceived, so far as regards the attributes of the Presiding Mind. For one can in imagination summon before us a small part[82] at least of the circumstances which must be contemplated and foreknown, before it can be decided what powers and qualities a new species must have in order to enable it to endure for a given time, and to play its part in due relation to all other beings destined to coexist with it, before it dies out. It might be necessary, perhaps, to be able to know the number by which each species would be represented in a given region 10,000 years hence, as much as for Babbage to find what would be the place of every wheel in his new calculating machine at each movement.

"It may be seen that unless some slight additional precaution be taken, the species about to be born would at a certain era be reduced to too low a number. There may be a thousand modes of ensuring its duration beyond that time; one, for example, may be the rendering it more prolific, but this would perhaps make it press too hard upon other species at other times. Now, if it be an insect it may be made in one of its transformations to resemble a dead stick, or a lichen, or a stone, so as to be less easily found by its enemies; or if this would make it too strong, an occasional variety of the species may have this advantage conferred upon it; or if this would be still too much, one sex of a certain variety. Probably there is scarcely a dash of colour on the wing or body, of which the choice would be quite arbitrary, or what might not affect its duration for thousands of years. I have been told that the leaf-like expansions of the abdomen and thighs of a certain Brazilian Mantis turn from green to yellow as autumn advances, together with the leaves of the plants among which it seeks for its prey. Now if species come in in succession, such contrivances must sometimes be made, and such relations predetermined between species, as the Mantis for example, and plants not then existing, but which it was foreseen would exist together with some particular climate at a given time. But I cannot do justice to this train of speculation in a letter, and will only say that it seems to me to offer a more beautiful subject for reasoning and reflecting on, than the notion of great batches of species all coming in, and afterwards going out at once."

Early in October Charles Darwin, for whose return from his noted voyage on the Beagle Lyell had more than once expressed an earnest desire, arrived in England, bringing with him a large collection of specimens and almost innumerable facts, geological and biological, the fruits of his travels. The biological observations slowly ripened in Darwin's mind till they had for their final result the "Origin of Species." The geological stirred Lyell to immediate enthusiasm, for they afforded a valuable support to some of the ideas which he had put forward to the "Principles." "The idea of the Pampas going up," he writes to Darwin, "at the rate of an inch a century, while the Western Coast and Andes rise many feet and unequally, has long been a dream of mine. What a splendid field you have to write upon!" The enthusiasm evidently was not confined to words, for Darwin himself says in writing to Professor Henslow, "Mr. Lyell has entered in the most good-natured manner, and almost without being asked, into all my plans."[83] The letter to Darwin,[84] which is quoted above, also contains a characteristic piece of advice.

"Don't accept any official scientific place if you can avoid it, and tell no one I gave you this advice, as they would all cry out against me as the preacher of anti-patriotic principles. I fought against the calamity of being President [of the Geological Society] as long as I could. All has gone on smoothly, and it has not cost me more time than I anticipated; but my question is, whether the time annihilated by learned bodies ('par les affaires administratives') is balanced by any good they do. Fancy exchanging Herschel at the Cape for Herschel as President of the Royal Society, which he so narrowly escaped being, and I voting for him too! I hope to be forgiven for that. At least, work as I did, exclusively for yourself and for Science for many years, and do not prematurely incur the honour or the penalty of official dignities. There are people who may be profitably employed in such duties, because they would not work if not so engaged."

Not very altruistic advice, it may be feared, but nevertheless bearing the stamp of practical wisdom. Committee-work and other official duties are terrible wasters of time, and thus, although often necessary and inevitable, are rightly regarded as evils. Many men, as Lyell intimates, have been seriously hindered in researches for which they were exceptionally fitted by allowing themselves to be at everyone's beck and call, and getting their days cut to shreds by meetings. So far has this gone in some cases, that the high promise of early days has been very inadequately fulfilled, and some great piece of work has been never completed. If the spirit in which Lyell writes were more frequent, the common illusion that workers in science belong to some inferior branch of the public service would be dispelled, and the business of scientific societies would sometimes run more smoothly; at any rate, it would be finished more quickly, because no one would care to waste time over splitting hairs, and hunting for knots in a bullrush.[85]

The year 1837, like the preceding one, was spent in quiet work, though three months of the summer were devoted to a journey on the Continent. As regards the former, it is evident that the book on which he was engaged had caused him more than ordinary difficulty, for it appears to have progressed more slowly than can be explained either by the duties of the Presidential chair, which he resigned in the month of February of this year, or by any distraction caused by other scientific work. But a sentence in a letter written to one of his sisters at the beginning of May throws some light on the cause of the delay. He says, "I have at last struck out a plan for the future splitting of the 'Principles' into 'Principles' and 'Elements' as two separate works, which pleases me very much, so now I shall get on rapidly."

The summer journey was to Denmark and the south of Norway, and this time Mrs. Lyell was able to bear him company. They left London early in June for Hamburg, crossing Holstein to Kiel, and travelling thence to Copenhagen. Here he set to work at once with Dr. Beck to study fossil shells, in the Crown Prince's cabinet and in the other museums of the city. Questions had arisen as to the nomenclature of various fossil species to which Lyell had referred in his book, on which Dr. Beck differed from Deshayes, so that Lyell was anxious to investigate some of the points for himself, and to see the original type-specimens in Linneus' collection, since these, in some cases, had been wrongly identified by Lamarck and other paleontologists. During a drive with the Crown Prince, he had the opportunity of examining an interesting section of the drift a few miles from Copenhagen, where it "was composed to a great depth of innumerable rolled blocks of chalk with a few of granite intermixed.

Fossils were numerous in the chalk.... Prince Christian set four men to work, while the horses were baiting, to clear away the talus, by which I saw that the boulders of chalk were in fact in beds, with occasional layers of sand between."

On reaching Norway Lyell made several expeditions from Christiania, in the course of which he examined a clay which occupies valleys and other parts of the granite region. This, which sometimes is found more than 600 feet above sea-level, he states "is a marine deposit containing recent species of shells, such as now inhabit the fjords of Norway."

This visit to Norway gave Lyell the opportunity of dispelling some erroneous ideas as to the relation of the granite to the "transition" (or lower Paleozoic) strata. This granite he found to be intrusive into these rocks, and into the much more ancient gneiss on which they rested. The sedimentary rocks near the junction were much altered, the limestones being changed into marble, the shale into micaceous schists; the fossils being more completely obliterated in the latter than in the former case. Some remarks which he makes as to the relations of the granite and gneiss indicate the closeness and carefulness of his observations. "This gneiss ... this most ancient rock is so beautifully soldered on to the granite, so nicely threaded by veins large and small, or in other cases so shades into the granite, that had you not known the immense difference in age, you would be half-staggered with the suspicion that all was made at one batch."[86]

From Copenhagen, on their return, they went to Leck and drove thence to Hamburg, across the sand and boulder formation of the Baltic, and so through the north of Germany. Among these boulders Lyell recognised the red granite, which he had seen in Norway sending off veins into the orthoceratite limestones and associated Silurian rocks. This "had been carried, with small gravel of the same, by ice of course, over the south of Norway, and thence down the south-west of Sweden, and all over Jutland and Holstein down to the Elbe, from whence they come to the Weser, and so to this or near this (Wesel-on-the-Rhine). But it is curious that about Munster and Osnabruck, the low Secondary mountains have stopped them; hills of chalk, Muschelkalk, old coal, etc., which rise a few hundred feet in general above the great plain of north and north-west Germany, effectually arrest their passage. This then was already dry land when Holstein, and all the Baltic as far as Osnabruck or the Teutoberger Waldhills, was submerged."[87]

At the end of September they returned to London through Paris and Normandy, and the rest of the year was mainly devoted to the completion of the "Elements of Geology." Little seems to have happened in the earlier part of the next year (1838); and in the summer Lyell went northward, halting on the way, at Newcastle-on-Tyne, to attend the meeting of the British Association. Here he was made President of the Geological Section, which appears to have been very successful, for he writes that the section was crowded--from 1,000 to 1,500 persons always present. The meeting, altogether, was a large one; but as the total number of tickets issued only amounted to 2,400, it seems probable that the general public was admitted more freely than is the custom at the present day. Sedgwick also on one occasion attracted a large crowd, for we are told that he delivered a most eloquent lecture "to 3,000 people on the Sea-shore." Geology, no doubt, has made great advances since that day, little more than half a century ago, but at the cost of much loss of attractiveness. It was then simple in its terminology, and fairly intelligible to people of ordinary education; now these are frightened away by papers bristling with technical terms and Greek-born words, and nothing but the prospect of a "scrimmage" would draw together 500 people to a meeting of Section C at the present day. Commonly the audience hardly amounts to one-fifth of that number. Geologists, perhaps, might consider with advantage whether a little abstinence from long words might not make the science more generally intelligible, and thus more attractive, without any loss of real precision.

The "Elements of Geology" was finally published a few weeks before the Newcastle meeting, and the work of recasting the "Principles" went on at intervals in preparation for the sixth edition, which appeared in 1840. If, in accordance with the maxim, a nation is happy which has no history, Lyell ought to have passed almost a year in a state of felicity, for nothing is recorded between September 6th, 1838, when he writes to Charles Darwin from Kinnordy, and August 1st, 1839, when he writes to Dr. Fitton from the same place. Both these letters are interesting. The former discusses the relation of Darwin's theory of the formation of coral islands with E. de Beaumont's idea of the contemporaneity of parallel mountain chains, which has been already mentioned. One passage also throws light upon the difficulties with which the British Association in its earlier days had to contend. Some of the most influential newspapers had set themselves to write it down--needless to say, without success. Good sense sometimes is too strong even for newspapers. But Lyell thus urges Darwin[88]:--

"Do not let Broderip, or the Times or the Age or John Bull, nor any papers, whether of saints or sinners, induce you to join in running down the British Association. I do not mean to insinuate that you ever did so, but I have myself often seen its faults in a strong light, and am aware of what may be urged against philosophers turning public orators, etc. But I am convinced--although it is not the way I love to spend my own time--that in this country no importance is attached to any body of men who do not make occasional demonstrations of their strength in public meetings. It is a country where, as Tom Moore justly complained, a most exaggerated importance is attached to the faculty of thinking on your legs, and where, as Dan O'Connell very well knows, nothing is to be got in the way of homage or influence, or even a fair share of power, without agitation."

Far-reaching words, the truth of which has been demonstrated again and again during the years which have elapsed since they were written. Lyell lays his finger on the weakest spot in the nature of the true-born Briton: he is deaf to quiet reasoning, and frightened by loud shoutings.

The second letter, that of 1839, is addressed to Dr. Fitton, who had written for the Edinburgh Review a criticism of the "Principles of Geology," in which he had expressed the opinion that Lyell had insufficiently acknowledged the value of Hutton's work. From this charge Lyell defends himself, pointing out that, valuable as were Hutton's contributions to the philosophy of geology, he was by no means the first in the field--that there were also "mighty men of old" to whom he felt bound to do justice, even at the risk of seeming to undervalue the great Scotchman. He points out that Hutton's work occupies a fair amount of space in the section of the "Principles" which is devoted to an historical sketch of the earlier geologists:--

"In my first chapter," he writes, "I gave Hutton credit for first separating geology from other sciences, and declaring it to have no concern with the origin of things;[89] and after rapidly discussing a great number of celebrated writers, I pause to give, comparatively speaking, full-length portraits of Werner and Hutton, giving the latter the decided palm of theoretical excellence, and alluding to the two grand points in which he advanced the science--first, the igneous origin of granite; secondly, that the so-called primitive rocks were altered strata.[90] I dwelt emphatically on the complete revolution brought about by his new views respecting granite, and entered fully on Playfair's illustrations and defence of Hutton.... The mottoes of my first two volumes were especially selected from Playfair's 'Huttonian Theory' because--although I was brought round slowly, against some of my early prejudices, to adopt Playfair's doctrines to the full extent--I was desirous to acknowledge his and Hutton's priority. And I have a letter of Basil Hall's, in which, after speaking of points in which Hutton approached nearer to my doctrines than his father, Sir James Hall, he comments on the manner in which my very title-page did homage to the Huttonians, and complimented me for thus disavowing all pretensions to be the originator of the theory of the adequacy of modern causes."[91]

In the following month Lyell attended a meeting of the British Association at Birmingham, and was invited, together with several of the leading men of science there present, to dine and spend the night at Drayton Manor, the residence of Sir R. Peel, near Tamworth. In a letter to one of his sisters, Lyell gives an interesting sketch of his impressions of the great statesman:--

"Some of the party said next day that Peel never gave an opinion for or against any point from extra-caution, but I really thought that he expressed himself as freely, even on subjects bordering on the political, as a well-bred man could do when talking to another with whose opinions he was unacquainted. He was very curious to know what Vernon Harcourt [the President for that year] had said on the connection of religion and science. I told him of it, and my own ideas, and in the middle of my strictures on the Dean of York's pamphlet[92] I exclaimed, 'By-the-bye, I have only just remembered that he is your brother-in-law.' He said, 'Yes, he is a clever man and a good writer, but if men will not read any one book written by scientific men on such a subject, they must take the consequences.' ... If I had not known Sir Robert's extensive acquirements, I should only have thought him an intelligent, well-informed country gentleman; not slow, but without any quickness, free from that kind of party feeling which prevents men from appreciating those who differ from them, taking pleasure in improvements, without enthusiasm, not capable of joining in a hearty laugh at a good joke, but cheerful, and not preventing Lord Northampton, Whewell, and others from making merry. He is without a tincture of science, and interested in it only so far as knowing its importance in the arts, and as a subject with which a large body of persons of talent are occupied."[93]

The next year (1840) appears to have slipped away uneventfully, for only a single letter serves as a record for the twelvemonth, and that is but a short one addressed to Babbage asking him to look up one or two geological matters during a journey through Normandy to Paris. As it is dated from London on the 11th of August, this looks as if Lyell did not go during the summer farther than Scotland, where he presided over the Geological Section at the meeting of the British Association.[94] The earlier part of 1841 appears to have been equally uneventful; but the summer of that year saw the beginning of a long journey and the opening of a new geological horizon, for Mr. and Mrs. Lyell crossed the Atlantic on a visit to Canada and the United States.

FOOTNOTES:

[66] At King's College and at the Royal Institution. See pp. 71, 72.

[67] Life, Letters, and Journals, vol. i. p. 401.

[68] Huxley, "Man's Place in Nature," p. 121.

[69] Only the skull was found, and that imperfect; moreover, the missing part could not be discovered. The same is true of the other animal remains, so that they could hardly have been victims of the Deluge.

[70] "Man's Place in Nature," p. 156.

[71] Turbo littoreus, Mytilus edulis, Cardium edule.

[72] The term, of course, is used here in the sense of either a slaty rock or a hard shale.

[73] The ruins of which (in the Bay of Bai? gradually sank after the middle of the fifth century until (probably towards the end of the fifteenth century) the floor was more than twenty feet under water. Since then it has risen up again.-

-"Principles of Geology," chap. xxx.

[74] He had expressed his doubts, in this and the former editions, as to the validity of the proofs of a gradual rise of land in Sweden.

[75] Life, Letters, and Journals, vol. i. p. 436.

[76] Lyell's specimens appear to have come from Kured, two miles north of Uddevalla, and only one hundred feet above the sea, but barnacles were obtained by Brongniart at two hundred feet.--"Principles of Geology," chap. xxxi.

[77] "Antiquity of Man," chap. iii.

[78] "Principles of Geology," ch. xxxi. "Antiquity of Man," ch. iii.

[79] "On the Structure of Large Mineral Masses," etc. Trans. Geol. Soc. Lond., iii. p. 461.

[80] Life, Letters, and Journals, vol. i. p. 460.

[81] The weakness of his eyes was always more or less of a trouble.

[82] It is "past" in the text (Life, Letters, and Journals, vol. i. p. 468), but I think this an obvious misprint.

[83] "Life of Charles Darwin," vol. i. p. 273.

[84] Life, Letters, and Journals, vol. i. p. 475.

[85] It is but rarely that, so far as the writer has seen, this remark applies to the committees of scientific societies in London, but the amount of time thus wasted in the universities, judging from his own experience of one of them, is really melancholy.

[86] Life, Letters, and Journals, vol. ii. p. 22.

[87] Ibid., vol. ii. p. 20.

[88] Life, Letters, and Journals, vol. ii. p. 45.

[89] Though undoubtedly this severance of geology and cosmogony was very helpful at the time to the progress of the former, the justice of it may be questioned; and Lyell's approval would not be endorsed by every geologist at the present day, though probably it would still commend itself to the majority.

[90] While this is true of many of the so-called primitive rocks, it is now generally believed that no inconsiderable portion are really abnormal or modified igneous rocks.

[91] Life, Letters, and Journals, vol. ii. p. 48.

[92] The Very Reverend W. Cockburn, D.D., who testified against the Association in a pamphlet entitled "The Dangers of Peripatetic Philosophy" (published in 1838). When the Association met at York in 1844, he read a paper before the Geological Section, criticising that science, and propounding a cosmogonical theory of his own. He was severely handled by Professor Sedgwick, but published his paper under the title, "The Bible defended against the British Association." This, though an exceptionally silly production, had a large sale. ("Life and Letters of Sedgwick," vol. ii. p. 76.)

[93] Life, Letters, and Journals, vol. ii. p. 51.

[94] Held at Glasgow, beginning September 17th. An allusion, however, during his American journey seems to imply a visit to France this year.

CHAPTER VII.

GEOLOGICAL WORK IN NORTH AMERICA.

This is a summary of their doings on the opposite side of the Atlantic in Lyell's own words: "In all, we were absent about thirteen months, less than one of them being spent on the ocean, nearly ten in active geological field work, and a little more than two in cities, during which I gave by invitation some geological lectures to large and most patient audiences."

To this may be added "three dozen boxes of specimens," and a mass of notes on the raised beaches of the Canadian lakes, the glacial drift, the falls of Niagara, and other questions of post-tertiary geology, as well as on the tertiary, cretaceous, coal, and older rocks. These afterwards produced a crop of about twenty papers, which appeared in various scientific periodicals. The principal results and the general impressions of the journey were worked up into a book entitled "Travels in North America," which was published in 1845.

A geologist who has been trained among the scenery of Britain finds his first view of the Alps to be the beginning of a new chapter in the Book of Nature, but a visit to America more like the beginning of a new volume. There almost everything is on a colossal scale--rivers, lakes, forests, prairies, distances, such as cannot be matched, at any rate in the more accessible parts of Europe. One may read of plains where the sun rises and sets as from a sea; of lakes, like Superior, as big as Ireland; of falls, like Niagara, where the neighbouring ground never ceases to quiver with the thud of the precipitated water; of rivers well nigh half a league wide while their waters still are far from the sea. But such things must be seen to be realised. In our own island Nature seems to be working at the present time on a scale comparatively puny; she must be watched as she puts forth her full strength before the adequacy of modern causes can be duly appreciated, and the history of the past can be understood by comparing it with that of the present.

The invitation to cross the Atlantic hardly could have reached Lyell at a more opportune epoch of his life. In his forty-fourth year, he was in full vigour both of mind and of body. A long course of study and of travel in Europe had trained him to be a keen observer, had enabled him to appreciate the significance of phenomena, and had supplied him with stores of knowledge on which he could draw for the interpretation of difficulties. America also offered a splendid field for work. Much of the country had been settled and brought under cultivation at no distant date; new tracts were being made accessible almost daily. Geologists of mark were few and far between, so that large areas awaited exploration, and in many places the traveller found a virgin field. The Geological Survey of Canada was just then being organised, the labours of the National Survey in the United States had not yet begun, though State surveys were at work, and had already borne good fruit. Indeed, while Lyell was in the country, the third meeting of the Association of American Geologists was held at Boston, and among those present were several men whose names will always occupy an honoured place in the history of the science. Still, at almost every step the observer might be rewarded by some discovery or by some fascinating problem which would give a direction to his future work.

The Lyells left Liverpool on July 20th, 1841, and reached Halifax on the 31st of the month, whence they went on to Boston, arriving there on August 2nd. The close resemblance of the shells scattered on the shore at the latter place to those in a similar situation in Britain was one of the first things which Lyell noted; for he found that about one-third were actually identical, a large number of the remainder being geographical representatives, and only a few affording characteristic or peculiar forms. For this correspondence, which, as he writes, had a geological significance, he was not prepared. The drifts around Boston, good sections of which had been exposed in making cuttings for railways, resembled very closely the deposits which he had seen in Scandinavia. Were it not, he says, for the distinctness of the plants and of the birds, he could have believed himself in Scotland, or in some part of Northern Europe. These masses of sand and pebbles, derived generally from the more immediate neighbourhood, though containing sometimes huge blocks which had travelled from great distances, occasionally exceeded 200 feet in depth. Commonly, however, they were only of a moderate thickness, and were found to rest upon polished and striated surfaces of granite, gneiss, and mica-schist. The latter effects, at any rate, would now be generally attributed to the action of land ice, but Lyell thought that the great extent of low country, remote from any high mountains, made this agent practically impossible, and supposed that the work both of transport and of attrition had been done during a period of submergence by floating ice and grounding bergs.

After a few days' halt at Boston, they moved on to Newhaven, where Professor Silliman showed him dykes and intrusive sheets of columnar greenstone altering red sandstone, their general appearance and association recalling Salisbury Crags and other familiar sections near Edinburgh. In this district Lyell found the grasshoppers as numerous and as noisy as in Italy, watched the fireflies sparkling in the darkness, and had his first sight of a humming-bird, and of a wildflower hardly less gorgeous, the scarlet lobelia.

From Newhaven they went to New York, and up the Hudson River in one of the great steamers, past the noble colonnade of basalt called the Palisades, and along the winding channel through the gneissic hills to Albany. Here a geological survey had been established by the State, and its members had already done good work, which, however, was not altogether welcome to its employers, for they had dispelled all hopes of finding coal within the limits of the State. This, as Lyell says, was a great disappointment to many; but it did good in checking the rashness of private speculation, and in preventing the waste of the large sums of money which had been annually squandered in trials to find coal in strata which really lay below the Carboniferous system. The advantage to the revenues of the state by the stoppage of this outlay and the more profitable direction given to private enterprise were sufficient, Lyell remarks, "to indemnify the country, on mere utilitarian grounds, for the sum of more than two hundred thousand dollars so munificently expended on geological investigation."

From Albany Lyell travelled to Niagara. The journey was planned in order to give him an opportunity of examining a connected series of formations from the base of the Paleozoic, where it rested on the ancient gneiss, to the coalfield of Pennsylvania; and he had the great advantage of being accompanied by one of the most eminent of American geologists, Mr. James Hall.

"In the course of this third tour," Lyell writes,[95] "I became convinced that we must turn to the New World if we want to see in perfection the oldest monuments of the earth's history, so far as relates to its earliest inhabitants. Certainly in no other country are these ancient strata developed on a grander scale, or more plentifully charged with fossils; and as they are nearly horizontal, the order of their relative position is always clear and unequivocal. They exhibit, moreover, in their range from the Hudson River to the Niagara some fine examples of the gradual manner in which certain sets of strata thin out when followed for hundreds of miles; while others, previously wanting, become intercalated in the series."

He observed, also, that while some species of the fossils contained in these rocks were common to both sides of the Atlantic, the majority were different; thus disproving the statement which at that time was often made--namely, that in the rocks older than the Carboniferous system the fossil fauna in different parts of the globe was almost everywhere the same, and showing that, "however close the present analogy of forms may be, there is evidence of the same law of variation in space as now prevails in the living creation."

Lyell made a thorough study of the Falls of Niagara, to which he paid a second visit before his return to England. The first view of these Falls, like the first sight of a great snow-clad peak, is one of those epochs of life of which the memory can never fade. It stirred Lyell to an unwonted enthusiasm. At the first view, from a distance of about three miles, with not a house in sight--it would be impossible, we think, to find such a spot now; "nothing but the greenwood, the falling water, and the white foam"--he thought the falls "more beautiful but less grand" than he had expected; but, after spending some days in the neighbourhood, now watching the river sweeping onwards to its final plunge, here in the turmoil of the rapids, there in its gliding, so smooth but so irresistible; now gazing at that mighty wall of 'shattered chrysoprase' and rainbow-tinted spray, which floats up like the steam of Etna; now looking down from the brink of the crags below the fall upon those rapids, where the billows of green water roll and plunge like the waves of the ocean, he "at last learned by degrees to comprehend the wonders of the scene, and to feel its full magnificence."

But, keenly as he might be impressed with the poetic grandeur of the falls, he could not forget the scientific questions which were ever present to his mind. The gorge of Niagara offered a problem for solution which had for him a special fascination. Not only did it illustrate on a grand scale the potencies of water in rapid motion, but also it furnished data for estimating the period during which this agent had been at work. The gorge has been carved in a plateau of Silurian rock, which terminates, seven miles below the falls, in a precipitous escarpment overhanging Queenstown. There was a time when that gorge did not exist, when the river first took its course along the plateau on its way from Lake Erie, and plunged over the brink of the escarpment. The valley at first was nothing more than a shallow trench excavated in the drift which covers the surface of the country--such an one as may still be seen between Lake Erie and the falls--but the river, slowly and steadily, has cut its way back through the rocky plateau from the first site of the falls near Queenstown to their present position. The upper part of this plateau consists of a thick bed of hard limestone, but beneath this the deposits become softer; and the lowest bed is the most perishable. The water, as it plunges down, undermines the overlying rock. The gorge began at once to be developed, and it has ever since continued to retreat towards Lake Erie. Every year makes some slight change. This becomes more marked when old histories are consulted and old drawings compared with the present aspect of the scene. Father Hennepin's sketch, of which Lyell gives a copy,[96] rude and incorrect as it is, proves beyond all question that the changes in the neighbourhood of Table Rock have been very considerable, for it shows that on this side a third and much narrower cascade fell athwart the general course of the main mass of water. This cascade, by the time of Kalm's[97] visit in 1751, had ceased to be conspicuous, and had quite disappeared before the date of Lyell's visit. The Horseshoe Fall also at the present time is less worthy of the name than it was at that date, for its symmetry has been seriously marred by a deep notch which the northern stream has cut in the more central part of the curve.[98] Careful inquiry convinced Lyell that the slow recession of the falls was an indubitable fact, and that its rate, on an average, was about a foot a year. As the gorge is about seven miles long, this would fix its beginning about 35,000 years ago.[99]

From Niagara Falls they travelled, still in Mr. Hall's company, by Buffalo to Geneva, examining on the way some red, green, and bluish-grey marls, with beds of gypsum and occasional salt springs, which, though older than the coal measures of England, closely resembled in appearance the upper part of the New Red Sandstone of Britain. Finally, after crossing the outcrops of the Devonian system, they reached Pennsylvania, where Lyell obtained his first view of the coal measures of North America, and was no less interested than surprised to find how closely the whole series corresponded with that of Britain. He saw sandstones "such as are used for building in Newcastle or Edinburgh, dark shales often full of ferns 'spread out as in a herbarium,' beds and nodules of clay-ironstone, seams of bituminous coal, varying in thickness from a few inches to some yards, and, beside these, an underlying coarse grit, passing down into a conglomerate, which was very like the millstone grit of England. The underclays beneath the seam of coal were full of stems and rootlets of Stigmaria, and the sight of these confirmed him in the opinion that the coal was formed of the remains of plants which had grown upon the spot."[100] After examining the district, they returned to Albany, and went thence to New York and Philadelphia, picking up on the way as much geological information as was possible.

New Jersey afforded some highly interesting sections of rocks belonging to the Cretaceous system, for these, though in mineral character resembling the greensands on the eastern side of the Atlantic, contained fossils which corresponded more closely with those of the white chalk, some species being actually identical. This fact was another proof that, though there had been in past ages a general similarity in the fauna of any period, geographical provinces had existed no less than they do at the present time.

Lyell had examined, as mentioned above, the bituminous coals in the undisturbed region of Pennsylvania, the next step was to study the beds of anthracite, with the associated strata, in the folded and broken ridges of the Alleghany Mountains. In this part of his work he had the inestimable advantage of being guided by Professor H. O. Rogers, whose name is inseparably connected with the geology of that classic region. The Alleghanies or Appalachians consist of a series of Silurian, Devonian, and Carboniferous strata in orderly sequence, "folded" (to use Lyell's words) "as if they had been subjected to a great lateral pressure when in a soft and yielding state, large portions having afterwards been removed by denudation. The long uniform, parallel ridges, with intervening valleys like so many gigantic wrinkles and furrows, are in close connection with the geological structure," and the rocks are most disturbed on the south-eastern flank of the chain, where the folds sometimes bend over to the west; in other words, the greatest disturbances are on the side nearest to the fundamental gneiss and the basin of the Atlantic--facts which probably stand in the relation of effect and cause.

It was a surprise to Lyell, on reaching the anthracite district around Pottsville on the Schuylkill, to see "a flourishing manufacturing town with the tall chimneys of a hundred furnaces, burning night and day, yet quite free from smoke." Special contrivances, of course, are requisite to secure the combustion of anthracite, especially in household fireplaces, but he had no hesitation in declaring that he preferred the use of it, notwithstanding the stove-like heat produced, to that of the bituminous coal consumed in London, with the penalty of living in an atmosphere dark with smoke and foul with smuts.

The seams of anthracite in this district are sometimes worked in open-air excavations, but as the strata have been bent into a vertical position the beds above and below, when the anthracite has been quarried out, are left like the walls of a fissure, and thus can be examined with the greatest ease.

Here also the "roof" of the seam proved to be a dark shale full of the usual plant-remains, among which were some British species of ferns, and the "floor" was an "underclay" containing the stems and rootlets of Stigmaria. Lyell also observed that the beds of detrital materials--sandstones, shales, etc.--were less persistent than those of coal, and that the way in which the former became thicker towards the south-east indicated that this was the direction of the ancient land region from which they had been derived. The result of his examination satisfied him that the anthracite of the Appalachians was identical in age, generally speaking, with the bituminous coal which he had previously examined, and was merely a fragment of the great continuous coalfield of Pennsylvania, Virginia, and Ohio, which lies about forty miles away to the westward.

After returning to Philadelphia Mr. and Mrs. Lyell went, vi?New York, to Boston, where he had been engaged to deliver a course of twelve lectures on geology at the Lowell Institute. To the courses here admission was free, but the tickets were given under certain restrictions. For Lyell's lectures about 4,500 were issued, and the class, he states, usually consisted of more than 3,000 persons. It had therefore to be sub-divided and each lecture to be repeated. The audience was composed "of persons of both sexes, of every station in society, from the most affluent and eminent in the various learned professions to the humblest mechanics, all well-dressed, and observing the utmost decorum."

At the conclusion of the lectures the Lyells travelled southwards, so that he might take advantage of the more genial climate and continue his geological work in the open air. He first halted at Richmond in Virginia, and from that place visited the Tertiary deposits in the vicinity of the James River. The more interesting of these are of Miocene age, and he observed that the fossils of Maryland and Virginia resembled those of Touraine and the neighbourhood of Bordeaux more closely than those from the coralline Crag of Suffolk, especially in the presence of genera indicative of a warm climate.

From this place they travelled across the "pine barrens"--where their train was stopped for the night by the slippery condition of the rails--to Weldon in North Carolina. Here Lyell saw the Great Dismal Swamp, a morass which extends for about forty miles from the neighbourhood of this town to Norfolk in Virginia. Like the bogs of Ireland, this marshy plain, some five-and-twenty miles across, is rather higher at the middle than at the edges. Its surface "is carpeted with mosses, and densely covered with ferns and reeds, above which many evergreen shrubs and trees flourish, especially the white cedar (Cupressus thyoides), which stands firmly supported by its long tap-roots in the softest parts of the quagmire. Over the whole, the deciduous cypress (Taxodium distichum) is seen to tower with its spreading top, in full leaf, in the season when the sun's rays are hottest, and when, if not interrupted by a screen of foliage, they might soon cause the fallen leaves and dead plants of the preceding autumn to decompose, instead of adding their contributions to the peaty mass. On the surface of the whole morass lie innumerable trunks of large and tall trees, blown down by the winds, while thousands of others are buried at various depths in the black mire below. They remind the geologist of the prostrate position of large stems of Sigillaria and Lepidodendron, converted into coal in ancient Carboniferous rocks."[101]

At Charleston they had practically passed beyond the southern limit of the winter snowfall, the greatest enemy of the field-geologist, and could carry on work without fear of interruption. Here they found flowers "at the end of December still lingering in the gardens," and were in the region of the palmetto palm. Few things during this rather lengthy journey impressed Lyell more than the facility of locomotion in a district which, comparatively speaking, was a new settlement, and was still in places thinly peopled, together with the general good quality of the accommodation for travellers. In this respect they had fared much worse during the previous year, when they were travelling through some of the more populous parts of France, such as Touraine and Brittany. After a journey through the pinewoods, they reached Augusta in Georgia, where another group of Tertiary deposits invited a halt. Those belonging to the Eocene period lie further down the Savannah River, so that a journey was made for the purpose of examining them, in the course of which, near the town of the same name as the river, Lyell also saw the clay in which remains of the mastodon and of other extinct mammals had been found. The muddy beach, with the tracks of racoons and opossums, gave him some hints as to the history of fossil footprints, so that on the whole very much interesting geology was the reward of a three weeks' stay in South Carolina. Then they once more turned their faces northward, and made their way, working at geology as they went, to Philadelphia, where they found themselves again in the region of colder winters at the present, and of erratic boulders as memorials of the past.

Six weeks were spent in Philadelphia, but Lyell's time was largely taken up by the delivery of a short course of lectures on geology. Pennsylvania, however, added to his experiences in another way, for the state had passed through a commercial crisis, and was unable to pay the interest on its funded debt. The soreness produced by this repudiation will not be readily forgotten, for nearly two-thirds of the stock--the whole amount of which was eight millions sterling--was held by British owners, so that the loss was felt heavily on this side of the Atlantic. In his "Travels" Lyell gives a brief history of this transaction, and discusses the political causes of a crisis which had been hardly less disastrous in America than in England.

They reached New York in the month of March, and spent several weeks there, for in that neighbourhood both the ancient crystalline rocks and the modern drift, with its erratics, afforded Lyell ample materials for study, each of these being then reckoned (and they have not ceased to be so counted) among the most difficult questions of geology. Towards the middle of April he proceeded northward, in order to examine the perplexing schists and less altered sedimentary deposits of the Taconic range, rocks which from that time to this have given ample employment to geologists. After this he found an opportunity of making use of the lessons learnt on the flats by the James River, for he went to Springfield and examined the famous footprints in the sandstone of Connecticut. As the deposit was referred to the Trias, and the footprints to birds, they were supposed to indicate the existence of this class of the animal kingdom at the beginning of the Secondary era. They have, however, now lost their special interest, since they are generally assigned to reptiles. After the middle of April was past, the travellers again reached Boston, from which city an excursion was made in order to study the Tertiary deposits of the island called Martha's Vineyard, off the coast of Massachusetts.

Returning to Philadelphia early in May, they went by Baltimore westward to the valley of the Ohio, in order to examine the undisturbed country beyond the folded district of the Alleghany Mountains. By this journey another section was, in fact, run across the great coalfield of the Eastern States, but considerably to the south of that which had been examined in the autumn of the preceding year. This proved no less interesting than the former one. At Brownsville, to take one instance only, a seam of bituminous coal, ten feet in thickness, was seen cropping out in the river cliff by the side of a large tributary of the Ohio, where it was worked by horizontal galleries. Pittsburg and other interesting localities in the neighbourhood were also visited, and then the Lyells descended the Ohio River to Cincinnati. He had thus traversed in descending order the succession of strata from the Carboniferous to the Lower Silurian or Ordovician system, which is exposed in the neighbourhood of that town. This, however, was not the only attraction offered by Cincinnati. Some two-and-twenty miles distant is the famous Big Bone Lick in Kentucky. Here some saline springs break out on a nearly level and boggy river plain, which are still attractive to wild animals, and often in past time lured them to their death in the adjacent quagmires. Here the bones of the mastodon and the elephant, of the megalonyx, stag, horse, and bison, have all been found, some in great numbers; and the last-named animals had frequented the springs within the memory of persons who were living at the time of Lyell's visit. These bones are generally embedded in a black mud, at a depth of about a dozen feet below the surface of the creek. Lyell suggests that very probably the heavy mastodons and elephants were lost by shoving one another off the tracks and into the more marshy ground as they struggled to satisfy themselves at the springs; just as horses, cattle, and deer get pushed into the stream in thronging to the rivers on the pampas of South America.

From Cincinnati the travellers struck northward to Cleveland on Lake Erie, going across a region which at that time was still being cleared and settled, and getting an experience of that American form of travellers' torture called a corduroy road. The lake-ridges--curious mounds or terraces of water-worn materials--in the neighbourhood of Cleveland afforded a new subject for an investigation which was continued in the vicinity of Ontario. But before reaching this lake Lyell spent a week at the Falls of Niagara, revising and enlarging the work already done. During the time he investigated the buried channel which appears to lead from the whirlpool to St. Davids, a league or so to the west of Queenstown. This was supposed by Lyell and many subsequent geologists to indicate part of an old course of the St. Lawrence, which had afterwards been blocked up by glacial drifts. It is, however, according to Professor J. W. Spencer, only a branch of a buried valley, outside the Niagara canyon and much shallower than it, which has been cut through by the present St. Lawrence, and has merely produced an elongation of the chasm at the Whirlpool.[102] Another series of lake-ridges was examined in the neighbourhood of Toronto. Here Lyell traced them to a height of 680 feet above the level of Ontario, seeing in all no less than eleven, some of them much reminding him of which he had examined in Sweden. In regard to these lake-ridges he writes thus:--

With the exception of the parallel roads or shelves of Glenroy and some neighbouring glens of the Western Highlands in Scotland, I never saw so remarkable an example of banks, terraces, and accumulations of stratified sand and gravel, maintaining, over wide areas, so perfect a horizontality, as in the district north of Toronto.[103]

Leaving Toronto on June 18th, they descended the St. Lawrence to Montreal and Quebec. The neighbourhood of either town afforded opportunities for much interesting work, especially in the drift deposits; the underlying ice-worn surfaces of crystalline or Paleozoic rock reminding Lyell of what he had seen in Scandinavia. At Montreal, the great hill, which gives its name to the town built upon its lower slopes, affords some highly interesting sections. It is composed of Paleozoic limestone, which has been pierced by more than one mass of coarsely crystalline intrusive rock and cleft by many dykes of a more compact character. Near the junction with the larger intrusive masses the limestone becomes conspicuously crystalline, and the fossils disappear, just as in the cases which Lyell had already seen about the border of granite in Scandinavia. Some also of the igneous rocks now possess a further interest, for they contain nepheline, a mineral not very common. This, however, had not been recognised at the time of Lyell's visit. The limestone in some of the quarries is wonderfully ice-worn, and the overlying drifts are in many ways remarkable. Of these drifts, Lyell examined various sections, at heights of from 60 to 200 feet above the St. Lawrence, finding plenty of sea-shells,[104] the common mussel being in one place especially abundant. He also examined some sections of stratified drifts between Montreal and Quebec, but without obtaining any fossils, though they had been found by Captain Bayford and others. The drifts, however, near the latter city were more prolific. With their shells, indeed, he was already, to some extent, familiar, for in the year 1835 he had received a collection from Captain Bayford. This happened to reach London at a time when Dr. Beck of Copenhagen was with him, and "great was our surprise," he writes, "on opening the box to find that nearly all the shells agreed specifically with fossils which, in the summer of the preceding year, I had obtained at Uddevalla in Sweden." The most abundant species were still living in northern seas, some in those of

Greenland and other high latitudes; while in Sweden they were found fossil between latitudes 58?and 60?N., and here in latitude 47? These fossil shells occur at Beaufort, about a league below Quebec, and about a quarter of a mile from the river, in deposits which have filled an old ravine in the Paleozoic rock. A laminated clay forms the lowest bed, above which comes a stratified sand, and this is followed by a clay containing boulders, each of these deposits being about twenty-five feet thick. They are without fossils, which begin with the next bed, a stratified mass of pebbly sand and loam, and become more frequent, till at last this passes into a mass nearly twelve feet thick, consisting almost wholly of the well-known bivalve Saxicava rugosa. This deposit was about 150 feet above the level of the sea. Afterwards, in travelling southwards from Montreal, whither he returned from Quebec, Lyell found marine shells on the border of Lake Champlain, about eighty miles from the former town. Here they occurred in a loam, which was covered by a sand, and rested on a clay about thirty feet thick, containing boulders, some of them nine feet in diameter.

Lyell sums up the results of his investigations by stating that, in his opinion, the shells certainly belong to the same geological period as do the boulders, and occur both above and below beds containing erratics; while the fundamental rocks below the drift are "smoothed and furrowed on the surface by glacial action." This effect Lyell at that time attributed to the friction of bergs grounding as they floated, but it is now referred by the majority of geologists to the action of land ice. Be this, however, as it may, the shell-bearing beds must have been deposited in the sea; so that either the land must have sunk as the ice retreated, or the latter at the time of its greatest extension must have trespassed on the domain of the sea, as it still does around parts of the Antarctic continent.

From Montreal they went, by way of Lake Champlain and over the Green Mountains, to Boston, where they arrived about the middle of July, and proceeded by steamer to Halifax. Here began the last stage of Lyell's journey, the examination of the Carboniferous system in Nova Scotia, to which work a full month was devoted. After studying the gypsum, red marl, and sandstone of the lower part of that system, which bears some resemblance to the Upper Trias (Keuper) of Britain, he crossed the Bay of Mines to Minudie, in the heart of the Nova Scotian coalfield. The cliffs by the sea-shore exhibit a fine series of sections, from the gypseous rocks up to the coal measures, uninterrupted by faults, the beds dipping steadily at an angle of nearly 30? Sandstones, shales, and seams of coal could be seen alternating in the usual manner; and from the last-named, stumps of trees, sometimes two or three yards high, were seen in places, as at South Joggins, projecting at right angles to the surface of the bed. Of such stems he observed at least seventeen at ten different levels. The stumps never pierced a coal-seam, but always terminated downwards either in it or in shale, and never in sandstone, thus indicating that they were a part of the vegetation from which the coal had been formed, and that it, like a peat-bog in England, required a subsoil impervious to water. Lyell also mentions that Mr. (now Sir) J. W. Dawson, who was his companion for part of the time, had found a bed of calamites in a similar position of growth.

But, in addition to much interesting work in various parts of the Nova Scotian coalfield, Lyell had the opportunity of witnessing the noted tides of the Bay of Fundy, where the difference between high and low water is as great as, if not greater than, anywhere else on the globe. On the muddy flats thus left bare he had another opportunity of studying the tracks left by various animals, marine and terrestrial; and in watching how these were hardened by the action of the sun, if they had been made near the high-water mark of spring-tides, he gained further hints for interpreting the fossil footprints of Connecticut and other countries.

On the 18th of August the Lyells left Halifax for England, thus bringing to a close a year of assiduous field-work, long journeys, and varied experiences. It was a period of the most continuous outdoor labour, and thus the most fruitful in the acquisition of knowledge which he had spent since his marriage and the publication of the "Principles of Geology"--a period comparable only with his journey, between May, 1828, and February, 1829, in France, Italy, and Sicily, though it was still longer and more fruitful, were this possible, in varied geological experiences. He had not, indeed, seen in this part of America any volcanoes, active or extinct--of which, however, he had already examined plenty; but he had studied good and characteristic sections of almost every formation which occurred in the more eastern states of America, from the most ancient crystalline masses, the foundation stones of the continent, to the most recent fossiliferous drifts. He had travelled from a region which resembled Scandinavia to one where the climate was more like that of the north coast of Africa, and had enlarged his conceptions of the scale on which Nature worked. But, in addition, he had been afforded an opportunity of studying the social and political condition of a young and vigorous nation as it was developing, unfettered by antiquated laws and hereditary customs. To this aspect of the tour a brief reference will be made in a later chapter; now it is enough to say that the long journeying of the twelvemonth had been happily ended, without illness, without the slightest accident, without anything that could be called an adventure. This good fortune followed them to the very end, for even the homeward passage is dismissed with the brief remark that it took nine days and sixteen hours; so that it may be supposed to have been prosperously uneventful. Then in eight hours after leaving Liverpool the travellers were back once more in London.

FOOTNOTES:

[95] "Travels in North America," chap. i.

[96] "Travels in North America," chap. ii.

[97] See the plate in the Gentleman's Magazine, 1751.

[98] See map in "Man and the Glacial Period," by Dr. G. F. Wright (International Scientific Series), p. 338.

[99] The estimates made by geologists have varied from 55,000 years (Ellicott, in 1790) to not more than 7,000 years (United States Geological Survey, 1886). Professor J. W. Spencer, who has recently investigated the question, has arrived, by a different method, at a date practically identical with that assigned by Lyell (Proc. Roy. Soc., vol. lvi. (1894), p. 145).

[100] This was still a moot point with geologists. Lyell refers to the confirmatory evidence which W. Logan had recently obtained in the South Wales coalfield of Britain.

[101] "Principles of Geology," chap. xliv.

[102] Proc. Roy. Soc. lvi. (1894), p. 146.

[103] The lake-ridges and raised beaches around the Great Lakes, indicating

margins of the water when it stood at a higher level than now, have received much attention of late years from Canadian and American geologists. They are found to vary somewhat in level, thus indicating unequal movements of the earth's crust. References to literature prior to 1890 will be found in a paper by Professor J. W. Spencer, Quart. Jour. Geol. Soc., vol. xlvi. (1890), p. 523.

[104] See, for descriptions of these sections and lists of the fossils, Sir W. Dawson's "The Ice Age in Canada," chaps. vi. and vii. They occur up to 560 feet above the sea.

CHAPTER VIII.

ANOTHER EPOCH OF WORK AND TRAVEL.

Very soon after their arrival in England the travellers went north to Kinnordy, where they remained till the end of October, when they returned again to their London home. Such an accumulation of specimens and of notes as had been gathered in America made necessary a long period of labour indoors, unpacking, classifying, and arranging; while certain groups of fossils had to be repacked and sent to friends, who had undertaken to work them out. These occupations apparently detained Lyell in London till August, 1843, when he started for Ireland, indulging himself on the way with a short run in Somersetshire for some geological work around Bath and Bristol, examining more particularly the "dolomitic conglomerate," a shore deposit of Keuper age, in which the remains of saurians had been found, and the Radstock Collieries, where he spent more than five hours underground "traversing miles of galleries in the coal," and finding here, as he had done in America, the stumps of trees in an upright position and shales full of fossil ferns as "roofs" to the seams. Then, in company with Mrs. Lyell, he crossed over to Cork, where the British Association assembled on August 17th, under the presidency of the late Earl of Rosse. The meeting was well attended by scientific men, but was coldly received by the neighbourhood and county-- partly, as Lyell says, because the gentry cared little for science; partly because the townspeople, comprising many rich merchants and most of the tradesmen, were "Repealers"; "and, the agitation having occurred since we were invited, the opposite parties could never, in Ireland, act or pull together."

It was impossible to visit Cork without seeing the beauties of the lakes and mountains of Killarney; and after this a short stay was made at Birr Castle, Lord Rosse's pleasant home at Parsonstown. The huge reflecting telescope, which is now more than a local wonder, was not then completed; but the smaller one, itself on a gigantic scale, was in full working order, and already had led to grand results by "not only reducing nebul?into clusters of distinct stars, but by showing that the regular geometric figures in which they presented themselves to Herschel, when viewed with a glass of less power, disappear and become very much like parts of the Milky Way." Thence they went northward to the coast of Antrim, to see the waves breaking upon the colonnades of basalt at the Giant's Causeway, and the dykes of that rock cutting through and altering the white chalk. Evidently the geology proved interesting, as well it might, for here Nature presents a volume of her geological history, that of the Secondary era, with only the opening and the concluding chapters, all the record from the early part of the Lias to the beginning of the Cretaceous having been torn out. The dark-tinted greensand, changing almost immediately into the pure white chalk, often presents curious colour-contrasts in a single section; while the classification of the several deposits offered a problem at which probably Lyell thought it wiser to "look and pass on." Several of the more interesting facts observed during this trip were afterwards described in the "Elements of Geology,"[105] among them the beds of lignite which occur in Antrim, associated with the great flows of basalt. Somewhat similar deposits were found, about seven years later, at Ardtun, in Mull, by the Duke of Argyll--a discovery which led Lyell to suggest, in later editions of the above-named work, the probability that the basalts of Antrim and of the Inner Hebrides were of the same geological age,--an inference which since then has been abundantly confirmed by the researches of Professor Judd and other geologists.

One of the most interesting sections in Scotland faces Antrim. Here, on the Ayrshire coast, between Girvan and Ballantrae, a complex of several kinds of igneous rock and a region, not a little disturbed, of "greywackes" and other sedimentary deposits present the geologist with problems more than sufficiently perplexing. At these Lyell took the opportunity of glancing, but a day's trip afforded no opportunity for any serious attempt to read the riddle. That had to be left to a later generation, and so it remained for over forty years. Something is now known about the igneous rocks, though here work still remains to be done; and the sedimentary deposits have been brought into order by the labours of Professor Lapworth. They exhibit, according to his description,[106] an ascending succession from the Llandeilo to the Llandovery group, and appear to be more modern than some, if not all, of the above-named igneous rocks. After their brief halt in this district the Lyells went on to Forfarshire, and spent the rest of the autumn at Kinnordy.

The winter was a busy time; he was writing steadily at his "Travels in North America," and working up some of the more distinctly scientific notes into formal papers for the Geological and other societies. Thus occupied, more than a year slipped away, diversified only by a summer visit to Scotland, attending the meeting of the British Association at York, and a journey to the Haswell Colliery, Durham, together with Faraday, as commissioners to examine into the cause of a recent disastrous explosion, and see whether such accidents could be prevented. Work at the "Travels in North America" took up all Lyell's spare time during the winter, and the book was published in the earlier part of 1845.

It was only a few months old when Mr. and Mrs. Lyell again set off for another tour in America. They left Liverpool on September 4th, and landed at Halifax on the 17th, after a voyage diversified agreeably by the sight of an iceberg and disagreeably by two gales. They went on at once to Boston, and thence made a tour through the State of Maine. During this sundry masses of drift were examined, which rested on polished and grooved surfaces of crystalline rock, and contained the usual shells, astarte, cardium, nucula, saxicava, etc., and in some places a fossil fish[107] in concretionary nodules. At Portland similar shells had been found in drifts which also contained bones both of the bison and of the walrus. These drifts in some places attained a thickness of 170 feet, and in them valleys 70 feet deep had been excavated by streams. Then they went to the White Mountains, and on approaching them Lyell did not fail to notice "on the low granite hills many angular fragments of that rock, fifteen to twenty feet in diameter, resting on heaps of sand." On their way they came to the Willey Slide, where a whole family of that name had been killed nineteen years previously in a landslip. Lyell carefully examined the scene of the accident, in order to ascertain what effects were produced by a mass of mud and stones as it slid over a face of rock, and found that it only made short scratches and grooves, not long and straight furrows, like those left by a glacier. They halted at Fabyan's Hotel near Mount Washington, and after waiting for a favourable day reached the summit (6,225 feet above the sea) on October 7th. It is easily accessible on horseback.

The notes of this excursion among the mountains show that Lyell still retained his old liking for natural history in general, for they contain remarks on the flowers, the insects, and the birds. Some observations on the Alpine flora of the higher summits in the White Mountains indicate his position at that time in regard to the origin of species. He adopts the hypothesis of 'specific centres,' viz. that "each species had its origin in a single birthplace and spread gradually from its original centre to all accessible spots, fit for its habitation, by means of the power of migration given it from the first." He supposed that the plants common to the more arctic regions and to the higher ground further south in Europe and Northern America were dispersed by floating ice during the glacial epoch, when the ground stood at a lower level, and that afterwards, when the climate became warmer, they gradually mounted up the slopes of the hills. The possibility of a migration by land is not mentioned, though doubtless it would have been admitted, because the evidence which he had so often studied pointed rather to a downward than to an upward movement but he asserts with some emphasis that many living species are older than the existing distribution of sea and land.

On his return to Boston, he had other opportunities of studying ice-worn rocks and erratics, and from this city made an excursion to Plymouth (Massachusetts) to see the spot where, on a mid-winter day, the Pilgrim Fathers had landed. But even here he could not neglect the shells upon the strand, and he records that eighteen species were collected, one-third of which were common to Europe. Still, we may note that on this journey rather more attention was paid than on the former to questions political, commercial, educational, and theological, and these occupy a larger space in the "Second Visit to the United States," which may account for its greater popularity. For example, it contains a sketch of the witch-finding mania in Massachusetts late in the seventeenth century, and a whole chapter on the sea-serpent. This "hardy perennial" had appeared in the Gulf of St. Lawrence in the previous August and in October, 1844,[108] and had repeatedly visited the New England coast from 1815 to 1825, when it had been seen by many credible witnesses. Lyell appears to be satisfied that, though allowance had to be made for exaggeration and honest misconception, some big creature had been seen, and suggests that it may have been an exceptionally large specimen of the basking shark.[109]

After a stay of nearly two months in Boston, they left for the south early in December, and found a little difficulty at first, as on a former occasion, from the slippery state of the rails. They journeyed by Newhaven, New York, Philadelphia, and Washington to Richmond, where a halt was made to examine the coalfield some sixteen miles to the south-west of the city. The measures rest on the granite, filling up inequalities on its surface, and are occasionally cut by dykes, which produce the usual alteration in the adjacent coal. The principal seam is from thirty to forty feet thick but the field, as a whole, reminded Lyell most of that at St. Etienne (France), which he had visited in 1843.[110] From Richmond they went, as on the former occasion, by Weldon to Wilmington, where the cliffs near the town yielded some Tertiary fossils, and on Christmas morning they landed from a steamer at Charleston.

From this city Lyell again visited the deposits near Savannah, which contained remains of megatherium, mastodon, and other large quadrupeds, as well as a second locality on Skiddaway Island, and then, on the last day of the year, quitted Charleston for Darien in Georgia. Here also were some more deposits of the same kind, while at St. Simon's Island Lyell examined a very large Indian mound. It was a mass of shells, chiefly of oysters, and contained flint arrow-heads, stone axes, and fragments of Indian pottery.

Returning to Savannah, they travelled towards the north-west, by Macon to Milledgeville. For more than 150 miles of the first part of the journey Lyell went along the railway on a hand-car, so as to study the cuttings and obtain the most continuous section possible of the Tertiary deposits from the sea to the inland granite. These deposits consisted of porcelain clays, yellow and white sands, and "burrstone," a flinty grit used for millstones, which often was full of silicified shells and corals, with the teeth of sharks and the bones of zeuglodon. Lyell mentions that in the neighbourhood of Macon he saw blockhouses such as those described by Cooper in the "Pathfinder," which twenty-five years earlier had been used for defence against the Indians before any white men's houses had been built in the forest.

Near Milledgeville the granite, gneiss, etc., is decomposed in situ to a considerable depth, and the rain-water, when the trees have been cut down, quickly furrows the detrital deposits of the neighbourhood. A remarkable instance of this action had occurred at Pomona Farm, where a ravine 180 feet broad and 55 feet deep had been excavated in the course of only twenty years.[111] From Milledgeville they returned to Macon, and thence travelled westward by Columbus to Montgomery, being much jolted in the stage-coach, but securing as a reward some Tertiary fossils; and at the latter place they found red clays and sandstones, which, however, were about the same age as the chalk of England. After the coach travelling, a journey by steamer down the Alabama River to Mobile was a welcome change, and the not unfrequent halts for cargo or to take in wood gave opportunities for collecting fossils from the neighbouring bluffs. One night they were startled by loud crashing noises and the sound of breaking glass, and found that the steamer had run foul of the trees growing on the bank. Their branches touched the water, as the river was unusually high; and the vessel, in the darkness, had been steered too near to the shore. Longer halts were made at Claiborne, to collect fossils from deposits corresponding in age with those at Bracklesham in England; and at Macon (Alabama) to visit a place where some remarkable specimens of the zeuglodon had been discovered. From Mobile also a long river journey was undertaken to Tuscaloosa, to visit a coalfield which supplied the town with fuel and the materials for gas. The field, "a southern prolongation of the great Appalachian coalfield," is a large one, being about ninety miles long and thirty wide, with some seams sixteen feet thick worked in open quarries. He remarks that he made geological excursions "through forests recently abandoned by the Indians, and where their paths may still be traced."

The strata on the Alabama River afforded a useful lesson on the variability of lithological characters. Were it not for the fossils, Lyell says, the Lower Cretaceous beds of loose gravel might be taken for the newest Tertiary, the main body of the Chalk for Lias, and the soft Tertiary limestone for the representative of the Chalk. It was impossible to leave Mobile without seeing something of the Gulf of Mexico; so they went in a steamer down the Alabama River to the seaside, looked upon the muddy banks, with the shells[112] which live in them and the quantities of drift-timber which bestrew them, and then went across to one of the minor mouths of the Mississippi, and, passing up it, landed at New Orleans.

This town, about 110 miles by water from the confluence of the main channel of the Mississippi with the sea, afforded a convenient opportunity for studying the character of the lower part of the delta of the "Father of Waters." Such a region might be expected to supply facts which would be helpful in the interpretation of many phenomena presented by the coal measures. Accordingly, Lyell made one excursion to Lake Pontchartrain, a great sheet of fresh water no great distance from both New Orleans and the sea, and another down to the mouth of the Mississippi. The road through the swamp to the former was constructed of a strange material--viz. the white valves of a freshwater mollusc.[113] These are obtained from a huge bank over a mile in length, and sometimes about four yards in depth, at one end of the lake. How this had been formed seemed doubtful. Possibly the shells had been piled up by the waves during a storm; possibly there had been some slight change of level. The lake itself is about fifteen feet below high-water mark, and is about as many deep; but, as it receives an arm of the Mississippi, silt is gradually raising the bottom. The sea sometimes, when impelled by a strong south-east wind, makes its way into the lake. Among the English coal measures--as, for instance, at Coalbrook Dale or in Yorkshire--beds of marine shells are occasionally found intercalated among or even associated with freshwater molluscs, without any alteration in the general character of the beds in which they lie. How this might occur is illustrated by Lake Pontchartrain in the swampy alluvial delta. Here a very slight physical change might enable the sea to take, for a time, possession of the land, and the denizens of its water, like a band of pirates, to dispossess the usual inhabitants.

The other expedition also supplied not a few valuable facts relating to the history of river deltas, which were afterwards supplemented as they travelled northwards for some hundreds of miles up the river, following its sinuous course through leagues of marshy plain, densely overgrown with vegetation. In the seaward reaches, reed, and rush, and willow, but above New Orleans cypresses and other timber trees, rise above the rank herbage.

The minor channels, blocked with driftwood which formed natural rafts; the sand-bars and mud-banks; the great curves of the river, the "bayous"[114] and isolated pools; the natural banks built up by the sediment arrested at flood-time by the herbage near the river brink; the floating timber and the "snags"--all provided valuable illustrations of the physical features of a great river delta, and supplied him with material which afterwards was worked up into newer editions of the "Principles" and the "Elements."

From New Orleans Lyell went by steamer to Natchez, halting on the way to examine more closely certain localities of interest and to obtain illustrations of how a coalfield might be formed. The bluffs of Natchez--almost the first place where distinctly higher ground approaches the river-side--afforded plenty of semi-fossil shells, specifically identical with those still inhabiting the valley of the Mississippi, but the loam in which they were embedded--a loam which reminded him of the loess of the Rhine--also contains the remains of the mastodon, and overlies a clay with bones of the megalonyx, horse, and other quadrupeds, mostly extinct. Beneath this clay are sands and gravel, the whole forming a platform which rises about 200 feet above the low river plain, revealing an earlier chapter in the history of the river. Similar bluffs occur at Vicksburg, but these disclosed Eocene strata beneath the alluvial deposits, and thus invited a halt in order to explore the neighbourhood. The next stage was to Memphis, nearly 400 miles. Lyell speaks highly of the accommodation generally afforded by the river steamers, but found the inquisitiveness of his American fellow-travellers rather a nuisance, and the spoiled children a still greater one. The former drawback to pleasure has certainly abated during the last half-century, but whether the latter has done the same may perhaps be disputed. New Madrid, 170 miles above Memphis, called for a longer halt, for the neighbouring district had suffered from a great earthquake in the year 1811, when shocks were felt at intervals for about three months, the ground was cracked, water mingled with sand was spouted out, yawning fissures opened (in one case draining a lake), portions of the river cliff were shaken down into the stream, and a large district--about 2,000 square miles in area--was permanently depressed. Some traces of the earthquake, in addition to the last-named, could still be recognised at the time of Lyell's visit, though more than thirty years had elapsed.

At Cairo, above New Madrid, the Ohio joins the Mississippi, and it was ascended to Mount Vernon. The geology now became a little more varied, for beneath the shelly loam already mentioned Carboniferous strata make their appearance, in which fossil plants are sometimes abundant and upright trees now and then occur. For nearly 200 miles higher up the Ohio, rocks of this age are exposed at intervals, till at last, near Louisville, those belonging to the Devonian system rise from beneath them. These, at New Albany, contain a fossil coral-reef, exposed in the bed of the river and crowded with specimens in unusually good preservation. At Cincinnati the travellers came at last upon old ground, and journeyed thence by steamer to Pittsburg. About thirty-two miles from this town, at a place called Greensburg, some remarkable footprints had been discovered on slabs of stone not many months before Lyell's visit, but as the beds on which they occurred belonged to the coal measures doubt had been expressed as to their being genuine, so he went thither to satisfy himself on this point. The footprints had disturbed the peace of Pittsburg, for they had started discussions in which one party had assumed, as matters of course, the high antiquity of the earth and the great changes in its living tenants, and had thus incurred the censure--which in some cases was followed by professional injury--not only of the multitude, but also of some of the Roman Catholic and Lutheran clergy. Commenting on this episode, Lyell quotes with approbation the words of a contemporary author,[115] which even at the present time occasionally need to be remembered:--"To nothing but error can any truth be dangerous; and I know not where else there is to be seen so altogether tragical a spectacle, as that religion should be found standing in the highways to say 'Let no man learn the simplest laws of the universe, lest they mislearn the highest. In the name of God the Maker, who said, and hourly yet says, "Let there be light," we command that you continue in darkness!'"

The travellers crossed the Alleghany Mountains in their way to Philadelphia. But a piece of work in Virginia had been left unfinished on the last occasion-- the examination of the Jurassic coalfield near Richmond. So he set off thither, leaving Mrs. Lyell in Philadelphia, and took the opportunity of examining the Tertiary deposits near the former town and the Eocene strata on the Potomac River. On his return they went to Burlington, which they reached in the first week in May, just as the humming-birds were arriving in hundreds, and by the 7th of the month they were in New York. The age of the so-called Taconic Group--a question of which so much has been heard of late years-- was then beginning to attract attention, so Lyell went in company with some American geologists to Albany in the hope of solving the problem. This he trusted he had done, but as his conclusions now would be deemed unsatisfactory, they need not be quoted. In reality, the question at that time was not even ripe for discussion.

On the homeward journey he turned aside at Boston to visit Wenham Lake, from which much ice was being supplied to London, and then they left for England by a steam packet which touched at Halifax. Four days after leaving this place they passed among a "group of icebergs several hundreds in number, varying in height from 100 to 200 feet," many of them picturesque in form, some even fantastic. Stones were resting on one of them, but as a rule they were perfectly clean and dazzlingly white, except on the wave-worn parts, which, as usual, were a beautiful blue. These, and a fine aurora borealis on the next night, were the only incidents of the voyage, and on June 13th, in twelve and a half days from Boston, the vessel reached Liverpool.

The close of this journey marks an epoch in Lyell's life. It was the last--unless we except his visit to Madeira--of his long wanderings for the purpose of questioning Nature face to face, and of studying her under various aspects and diverse conditions. He did not, indeed, cease to travel. He twice returned to America, he revisited Sicily and various parts of Europe, but these journeys not only occupied less time but also led him among scenes for the most part not unfamiliar. He doubtless felt that on reaching his fiftieth year he might fairly regard the more laborious part of his education completed, although he never ceased to be a learner, even to the latest days of his life, when strength had failed and memory was becoming weak.

An account of the above-named journey was published in 1849, under the title of "A Second Visit to the United States of North America." This book, in addition to descriptions of the scenery and the geology of the country, contains much general information about the people, with remarks by the author on various political questions, such as the condition of parties, the effects of almost universal suffrage, particularly on the national sense of honour and morality, the existence and evils of slavery, the state of religious feeling, the position of Churches, and the systems of education, especially when contrasted with those of England. Some of these questions about this time were exciting much attention in Great Britain, and in regard to one matter--the delimitation of the territories of the two nations in the region west of the Rocky Mountains--friction existed, which was so serious that more than once war seemed possible. On this account, probably, the "Second Visit" was a greater success, commercially speaking, than the "Travels," for it reached a third edition.

FOOTNOTES:

[105] Chapters xiv. and xxix.

[106] "The Girvan Succession," Quart. Jour. Geol. Soc., xxxviii. (1882), p. 537.

[107] The capelin (Mallotus villosus), which still lives in the Atlantic.

[108] It was also seen the following year on the coast of Virginia, and on that of Norway in both 1845 and 1846.

[109] He says that the alleged sea-serpent washed ashore at Stronsa (Orkneys) in 1808 is proved by the bones (some of which are preserved) to have been this animal.

[110] The formation, however, does not belong to the Carboniferous system, but is shown by its fossils to be Jurassic in age.

[111] It is described and figured in later editions of the "Principles of Geology," chap. xv. (eleventh edition).

[112] A species of Gnathodon.

[113] Gnathodon cuneatus.

[114] A bayou is the name given to an old channel of the river. When the latter is making a series of horseshoe curves, the stream often cuts through the neck of land which separates its nearest parts. The water then takes the shortest course, the entrances to the old channel are silted up, and it becomes a horseshoe-shaped pool.

[115] T. Carlyle ("Letter on Secular Education").

CHAPTER IX.

STEADY PROGRESS.

The "Principles of Geology" had been completed and published for thirteen years, yet catastrophism, as we learn from a correspondence with Edward

Forbes,[116] dated September, 1846, was dying hard. "Agassiz, Alcide D'Orbigny, and their followers [were still] trying to make out sudden revolutions in organic life in support of equally hypothetical catastrophes in the physical history of the globe."[117] A remark in Forbes's reply is striking:--

"You are pleased to compliment my paper on its originality. Any praise from you must ever be among the greatest gratifications to me, and to any honest labourer in the great field of Nature. But I had rather hear the views I have set forward be proved not original than the contrary. It seems to me that the surest proof of the truth of such conclusions as I have summed up at the end of my essay is the fact of their not being original so far as one person is concerned, and of their having become manifest to more than one mind, either about the same time or successively, without communication. I believe laws discover themselves to individuals, and not that individuals discover laws. If a law have truth in it, many will see it about the same time."

In this month also the Lyells removed from Hart Street to 11, Harley Street. The house where they had spent fourteen years very happily was not left without regret, but it had become too small. They had no children, but a rapidly increasing geological collection takes up almost as much room as (though it is much more silent than) a growing family. The removal of a geological collection is a laborious business; and, besides this, Lyell was preparing a new edition of the "Principles" and writing a book about his recent travels in America. Still, to judge from his letters, he found time for some pleasant social distractions; for his letters to the old home at Kinnordy contain more often than formerly interesting references to talks with such men as Macaulay, Milman, and Rogers, Lord Clarendon and Lord Lansdowne. The seventh edition of the "Principles," condensed into a bulky single volume, was published early in 1847, and in the following June Lyell attended the meeting of the British Association at Oxford, which appears to have been no less pleasant than successful, although "out of twenty-four Heads of Houses only four were at Oxford to receive the Association." On this occasion, he writes, he became better acquainted with "Ruskin, who was secretary of our Geological Section." The remainder of this summer was spent in Scotland, and the rest of the year, with most of the following one, was devoted to quiet work. Still, Lyell took an active part in a crisis through which, about this time, the Royal Society was passing. A number of the Fellows, including most of those eminent in science, were anxious to raise the standard for admission into the Society. For many years past the "three letters" had often signified little more than an indication of good means and social position, coupled with a certain interest in scientific pursuits. The reformers prevailed, after a long struggle "with a set of obstructives compared with whom Metternich was a progressive animal," and the present status of the society is the result. Incidental remarks in Lyell's letters to his relations also indicate that he was becoming well known in circles other than scientific, of which a further proof was given in the autumn of 1848, when he received the offer of knighthood. Of course, in any country where "orders of merit" exist, other than Great Britain, Lyell would have been "decorated" years ago, but we manage things differently. As a rule, we let science and literature be their own reward, and, as an exception, confer the same distinction on a man who has won a worldwide reputation (provided he is fairly rich) and on an opulent tradesman who is accidently prominent on some auspicious occasion, or is a local wirepuller in party politics. Lyell went over from Kinnordy to Balmoral to receive the intended honour, and had, as he writes, "a most agreeable geological exploring on the banks of the Dee, into which Prince Albert entered with much spirit." In February, 1849, he was elected for the second time President of the Geological Society, and in the autumn, when at Kinnordy, was again invited to Balmoral, where he had some interesting talks with Prince Albert on subjects ranging from various educational and broad political questions to the entomology of Switzerland, Scotland, and the Isle of Wight.

In the middle of September he attended the meeting of the British Association at Birmingham, where he was for the third time President of the Geological Section. A few weeks later his father, whose health had been for some time failing, died at Kinnordy.[118] The latter was a rich man, but as he made liberal provision for his daughters and younger sons, Sir Charles, though he succeeded to a considerable estate, found himself unable to afford the expense of keeping up Kinnordy as well as a house in London. Which, then, was henceforth to be his home? The attractions of Kinnordy were obvious, but the long distance from the metropolis was a serious drawback, while the duties of a resident landlord would have interfered much with his geological work, which would have been still more hampered by the severance from libraries, museums, and intercourse with fellow-workers. Thus he felt it his duty to retain his house in London and to let Kinnordy, though, as his mother and sisters retreated to the "dower house," he was able from time to time to visit the old place. The decision probably was less painful than it otherwise would have been from the fact that his boyhood had been spent in England. At any rate, it was a wise one, in regard to both his own reputation and the progress of science in general.

In the summer of 1850, Sir Charles augmented his experience and refreshed old memories by a tour in Germany. During this he saw for the first time the Roth-todt-liegende or Lower Permian conglomerates at Halle and at Eisenach, as well as the great lava streams which had supplied them with so much of their materials. Also he went to the Brocken in order to examine into Von Buch's extraordinary assertion that the granite had "come up in a bubble." This, it is needless to say, was speedily pricked. The loess also, that singular deposit which wraps like a mantle so much of the undulating ground in Northern Germany, evidently engaged his attention, and we find the fruits of these studies in a later work. In addition to all this, he did more than glance at the Maestricht Chalk, the "Wealden" coal of Hanover, the Tertiary deposits near Berlin, the Paleozoic rocks of the Hartz, and the scenery of the Saxon Switzerland.

His books, his scientific papers, and Presidential addresses to the Geological Society, his duties as a commissioner, at first for the Exhibition of 1851, and somewhat later for the reform of the University of Oxford, kept him pretty well employed till August, 1852, when he for the third time crossed the Atlantic to deliver another course of lectures at the Lowell Institute, Boston. Though he was back in England before Christmas, he found time for some geological work in America, the most important item in which was an excursion from Halifax in company with his old acquaintance, Mr. J. W. Dawson, to the Nova Scotian coalfield. On this occasion he passed through a fair amount of country still uncleared, which made the journey more interesting; he had also opportunities of appreciating the effects of ice in moving and piling up boulders on the shores of lakes, and obtained still more evidence in regard to this, on reaching the sea-coast in the neighbourhood of the coalfield. But their labour was rewarded by one discovery of exceptional importance. In the trunk of a tree which had died and become hollow in a forest of the Carboniferous period, they found entombed the skeleton of an animal. Whether this were a fish or a reptile was at first hotly disputed, but finally it proved to be an amphibian.

On his return to England, Sir Charles was kept for some time fully employed by the preparation of the ninth edition of the "Principles," but early in the summer of 1853 he went for the fourth time to America--on this occasion in company with Lord Ellesmere--as commissioner to the Exhibition held at New York. But now his time was fully taken up by official duties, and his visit was a short one, for he returned before the end of July, and was soon afterwards invited to visit Osborne and give some account of his journey to the Queen and Prince Albert.

Very early in 1854 he again left England, in company with Lady Lyell and Mr. and Mrs. Bunbury, to visit Madeira. Some three weeks were devoted to a careful study of the geology of that island,[119] partly with the view of determining whether it afforded any support to Von Buch's favourite notion that volcanic cones were mainly formed by upheaval. As might be anticipated, the evidence was distinctly unfavourable. The island was proved to be mainly composed of volcanic material, cones of basaltic scoria, and great flows of similar lava, which had been piled successively one on another in the open air to a depth of about 4,000 feet. This mass had been subsequently pierced by dykes, worn by storm and stream, and in one or two places deeply grooved by rivers. There were, indeed, some underlying beds of marine origin, which, in one part of the island, rose to a height of 1,200 feet above the sea, and thus indicated a certain amount of upheaval; but even this was not of the kind which Von Buch's hypothesis required, while the rest of the evidence, including that afforded by some tuffs containing fossil plants, proved that the major part of the island had been formed above water.

From Madeira they went on to Teneriffe, Palma, and the Grand Canary. Of this part of the journey few details are given, but the results were afterwards incorporated with one of his books.[120] To the Peak of Teneriffe the reference is comparatively brief. Of Palma the account is much fuller, for this island had been regarded by Von Buch, who visited it in 1825, as a type of his "craters of elevation"--an idea which was dispelled by Lyell's investigation. The Grand Canary, like Madeira, proved to be formed of masses of subaerial volcanic rock, perhaps even thicker than those in Madeira, which also rested upon some upraised marine deposits of Miocene age.

In the course of 1854 Sir Charles received from his own University the honorary degree of D.C.L. Much time was spent in working up the results of his last journey, some of which were communicated to the Geological

Society.[121] In the spring of 1855 he went to the Continent, studying, among other matters, the drifts in the neighbourhood of Berlin. In the summer he visited Scotland, made the acquaintance of Hugh Miller, worked over Arthur's Seat, Blackford Hill, and "the coast of Fife from Kinghorn to Kirkcaldy." It would be hard to find a set of sections better adapted for the study of ancient volcanic rocks, both contemporaneous and intrusive, than this coast affords; and his experience in Madeira and the Canaries enabled him to regard "the Edinburgh and Fife rocks with very different eyes."

One or two of his published letters about this period have a special interest, for they show that his views on the origin of species were undergoing a gradual modification. Speaking of some strange variations in the flower of an orchideous plant,[122] he refers, half in jest, to "ugly facts, as Hooker, clinging (like me) to the orthodox faith, calls these and other abnormal vagaries"; and again, the following sentences do not come from a man who is firm in his belief[123]:--

"When Huxley, Hooker, and Wollaston were at Darwin's last week, they (all four of them) ran a tilt against species further, I believe, than they are deliberately prepared to go--Wollaston least unorthodox. I cannot easily see how they can go so far, and not embrace the whole Lamarckian doctrine. Huxley held forth last week about the oxlip, which he says is unknown on the Continent. If we had met with it in Madeira and nowhere else, or the cowslip, should we not have voted them true species? Darwin finds, among his fifteen varieties of the common pigeon, three good genera and about fifteen good species, according to the received mode of species and genus-making of the best ornithologists, and the bony skeleton varying with the rest! After all, did we not come from an ourang, seeing that man is of the Old World, and not from the American type of anthropomorphous mammalia?"

Sir Charles and Lady Lyell were again on the Continent in the summer of 1856, examining the drifts of Northern Germany, visiting Humboldt at Berlin, discussing geological questions, especially in regard to Carboniferous plants, at Breslau with Roemer and Goeppert; working over the Riesengebirge; then going on to Dresden, and passing through the Saxon Switzerland to Aussig. The coalfield north-west of the former city was not neglected, the great breccia beds of the Rothliegende were again examined, and account was taken of Ramsay's opinion that certain British Permian breccias were glacial in origin. Close attention was also bestowed upon the great masses of hard quartzose grit, through which the Elbe has carved its way--the Quader of Saxony; for this formation, "a grit wholly deficient in calcareous matter, corresponds to the more purely calcareous rock (Chalk) of Great Britain, and yet contains here and there the same shells." He did not neglect the Brown Coal[124] between Tlitz and Aussig, and, on reaching Prague, made the acquaintance of Barrande, who took him to see those older Paleozoic rocks among which the great paleontologist had been labouring for nearly a quarter of a century. Then the travellers proceeded to Vienna, and after that to the Styrian Alps, to visit various interesting sections in the Salzkammergut, such as the classic ground at Gosau and the Triassic limestones near Hallstadt, where the last survivors of the Paleozoic ages are entombed with the representatives of the period. His letters, like many others of earlier date, indicate that, notwithstanding the fascinations of geology, neither living molluscs, nor insects, nor plants had ceased to interest. They returned by way of Munich, Ulm, Zurich and Paris, reaching England about the end of October.

The summer of 1857 was devoted to another Continental tour, rather more restricted than the former, but by no means unimportant. They went leisurely through Belgium and up the Rhine into Switzerland, halting at different places either to study sections of special interest or to confer with eminent geologists. Part of a letter written at this time[125] gives a valuable insight into the intention of these journeys and the character of the author, who was now in his sixtieth year:--

"I hope to continue for years travelling, making original observations, and, above all, going to school to the younger, but not, for all that, young geologists, whom I meet everywhere, so far ahead of us old stagers that they are familiar with branches of the science, fast rising into importance, which were not thought of when I first began."

Switzerland, obviously, was visited on this occasion with a very definite purpose. De Charpentier, Escher von der Linth, and other local geologists, had been for some time asserting that the glaciers of the Alps, at no remote epoch in geological history, had attained to an enormous size, had buried the Swiss lowland and covered it with morainic deposits, and had even welled up high against the flanks of the Jura, where the huge blocks of protogine from the Mont Blanc range--such as Pierre ?bot and its companion erratics, full 800 feet above the Lake of Neuchâtel--indicated one position of its terminal moraine. Formerly, in common with many other geologists, Sir Charles had supposed these blocks to have been transported from the Alpine peaks by ice-rafts on the sea, at a time when the whole region stood at a considerably lower level. But now, after examining the erratics, their regular and significant distribution, the other glacial débris, the ice-worn surfaces of rock beneath it, and ascertaining the distinctly terrestrial character of the deposits all about the mountains, he unreservedly admitted land-ice to be the only possible agent, and, in accepting this hypothesis, perceived clearly that he must not shrink from applying it to Scotland. Then he plunged into the mountains to examine and follow the track of the retreating ice-sheet up to the glaciers which are still at work among the higher peaks, passing up the valley of the Reuss, crossing the Furka Pass, and descending the Rhone valley to Visp, but turning aside to examine the earth pillars on the flank of the Eggishorn.[126] Another, and a larger group of these pillars--instances of the erosive action of rain-water on morainic material--was seen near Stalden, in the Visp-thal; but these had been damaged by the earthquake which two years before had severely shaken this part of the Alps. At Zermatt the characteristics of glaciers and the effects of ice were carefully studied among the grandest of Alpine scenery; then, on returning to the Rhone Valley, they crossed the Alps by the Simplon and went on to Turin. Here he took the opportunity of visiting the huge moraine near Ivrea, which rises from the lowland like a range of hills, and of investigating the erratics of the Superga, satisfying himself that they really belonged to the Miocene deposits of that hill, and were indicative of the existence of glaciers in the Alps of that epoch, which had been large enough to reach the sea-level, and to send off masses of ice laden with boulders. Then they went on to Genoa, and along the beautiful Riviera di Levante to Pisa; thence, after a short visit to Florence, proceeding direct from Leghorn to Naples. Here, he once more examined Vesuvius, and had the luck to see lava streams actually in motion--"some going fast, others going very slow"--a sight which "gave him many new ideas." A study also of the dykes of Somma convinced him that they afforded no support to De Beaumont's idea of a distension of the mass.[127]

From Naples he went to Sicily, in order to make a second examination of Etna, and then, after rejoining Lady Lyell, spent some time in the neighbourhood of Rome, visiting the old volcanic district of the Alban Hills, and making excursions, as they travelled northward, into the Apennines. They returned through France, reaching London towards the end of December.

But, for a worker so thorough in his methods, this visit to the volcanoes was not enough, so next year, after spending the earlier part of the summer with his brother's[128] family in the neighbourhood of Darmstadt, he left Lady Lyell there, and set off towards the end of August for a third examination both of Vesuvius and of Etna. Travelling rapidly up the valley of the Rhine, he went by Geneva to Culoz, and over Mont Cenis to Turin and Genoa, without halting for geological work, and thence by sea to Naples. Lava was still flowing from Vesuvius, that black mass, with its strange rope-like folds and slaggy wrinkles,[129] now so well known to every visitor. Accompanied by Professor Guiscardi--one of the most genial and helpful of leaders--Sir Charles made his way to a vent at the base of the principal cone, where the lava was still welling forth from "a small grotto, looking as fluid as water where it first issued, and moving at a pace which you would call rapid in a river. White-hot, at first, in a canal four or five feet broad, then red before it had got on a yard, then in a few feet beginning to be covered by a dark scum, which thickened fast and was carried along on the surface." But the great question, whether a volcano was mainly a "crater of elevation" or a "crater of ejection," was ever present to his mind; so, in addition to studying the grand sections displayed in the crags of Monte Somma, he devoted two days to the exploration of the ravines which furrow its outer slopes. He also found time to have another look at the Temple of Serapis, and to examine the Solfatara, which is a striking example of a crater at once broad and low.

After a week's halt at Naples, Sir Charles resumed his journey to Sicily, landing at Messina on September 10th. By the 15th he was once more on the slopes of Etna, and had begun a twelve-day period of hard work on the mountain, passing five nights in very rough quarters at the Casa degli Inglesi, 9,600 feet above sea-level. During this stay he ascended the principal cone, carefully examining both the larger and the smaller craters, and descended into the Val del Bove, a laborious expedition, but one which well repaid him by throwing much light on the structure of the volcanic mass. Still he was not yet satisfied, for after he had descended to Zafarana, he returned to spend another night at the Casa degli Inglesi in order to satisfy himself about one or two details. From Zafarana also he went again to the Val del Bove, checking and increasing his notes, and devoted another day to a most interesting excursion through picturesque scenery as far as the watershed between this vast hollow in the mountain side and the neighbouring Val di Tripodo. On all these excursions Sir Charles, as far as possible, rode, remarking to his wife, "I feel here that a good mule is like presenting an old geologist with a young pair of legs." Work on the mountain ended, he spent a little time in examining the Tertiary beds of the neighbouring lowland, and then, getting back to Messina about the middle of October, returned in due course to England.

These two journeys in succession greatly augmented his knowledge of the structure of volcanic cones, and enabled him to deal the death-blow to the "crater of elevation" hypothesis which had found such favour among Continental geologists. He could now prove that lava would solidify in a compact form on slopes of thirty-five or even forty degrees--a fact which had been stoutly denied by advocates of that hypothesis, and was able to offer an explanation of the singular structure of the Val del Bove, viz. that it was a huge gulf, formed by a series of mighty explosions, similar to those which shattered half of the old crater of Vesuvius,[130] and sent one side of Bandai San[131] flying through the air. He returned to England satisfied that his feet were on firm ground, if such a phrase be permissible in regard to a volcano, and that the results[132] of this conscientious labour in the fulness of his age had strengthened him in the position which he had adopted in his scientific youth.

In the next year (1859) Lyell also travelled, though the journeys were not so lengthy as their two predecessors. Still, in the spring he visited both Holland and Le Puy in Auvergne, and in the earlier part of the autumn attended the meeting of the British Association at Aberdeen, under the presidency of Prince Albert. A strong body of geologists were present, and Lyell was for the fourth time in the chair of the Geological Section, the Prince coming to hear his address. Among the old friends whom he met was one who would have been a suitable husband for the famous Countess of Desmond, for Lyell writes of him to Mrs. Horner, his wife's mother, "Dr. F. at ninety-four looks well enough, but having eaten turtle-soup, and melon too close to the rind, and other imprudences, is not quite well to-day!" O dura Doctorum ilia! The meeting ended, Lyell with some geological friends went off to Elgin to examine the sandstone quarried at Cutties Hillock, near that town. The rock closely resembles the ordinary Old Red Sandstone; it seemed at first sight to form a continuous mass, yet in one place it contained a fossil fish belonging to that period, and in another the remains of a reptile (Telerpeton). After some days of careful study, the Rev. W. S. Symonds, who was one of the party, came to the conclusion (which has been fully ratified by later investigations) that the deposits were of different ages; the one with the fish being truly "Old Red," the other, with the reptile, "New Red." The chief cause of the puzzle is that the sand which has been derived from the older rock has gone to form the newer one, and that the usual indications of a discontinuity are practically absent. It affords a valuable caution, for it shows that Nature sometimes does set traps, which might well catch even the most wary geologist.

In the same autumn Lyell read Darwin's great work on "The Origin of Species," by which his scientific position was finally determined, for his letters show that, if any objection to the leading principles in his friend's views had still lingered in his mind, they were overcome by the perusal of this masterly specimen "of close reasoning and long sustained argument."

FOOTNOTES:

[116] In reference to an essay written by him on the connection between the fauna and flora of the British Isles and geological changes. ("Memoirs of the Geological Survey," i. p. 336.)

[117] Life, Letters, and Journals, vol ii. p. 110.

[118] He died November 8th, 1849.

[119] He had the advantage of the company of Mr. C. Hartung, who was an excellent naturalist and well acquainted with the island.

[120] "Elements of Geology" (sixth edition), pp. 621-635.

[121] "On the Geology of Some Parts of Madeira" (Quart. Jour. Geol. Soc., x. p. 325).

[122] In a letter to Mr. Bunbury, dated November 13th, 1854 (Life, Letters, and Journals, vol. ii. p. 199). It is written from 53, Harley Street, one in the previous August bearing the superscription of 11, Harley Street, so that he appears (though there is no allusion to this in his published letters or journals)

to have removed into another house in the same street. The number of this was subsequently altered.

[123] Another letter to Mr. Bunbury, dated April 30th, 1856 (ibid., p. 212).

[124] This deposit belongs to the Tertiary era (Oligocene system).

[125] Life, Letters, and Journals, ii. p. 243.

[126] The largest, called the Zwerglithurn, is about one and a half hours walk above Viesch.

[127] This had been asserted in support of the hypothesis of "craters of elevation."

[128] Colonel Lyell had retired from the army and returned to England a short time before the outbreak of the Indian Mutiny.

[129] See Professor J. W. Judd: "Volcanoes" (International Scientific Series), Fig. 22.

[130] In the famous eruption of A.D. 79.

[131] A volcano of Japan.

[132] These results are worked into the tenth edition of the "Principles" (chaps. xxv. and xxvi.). See also a paper on Stony Lava on Steep Slopes of Etna (Proc. Roy. Soc. 1858, ix. p. 248). He received the Copley Medal from the Royal Society in November.

CHAPTER X.

THE ANTIQUITY OF MAN.

Though many men on reaching their sixty-third year are content to rest upon their oars and not to attempt new ventures, Lyell had plunged into a question which was arousing almost as much excitement as the origin of species--namely, the antiquity of man. It was a question, indeed, which for a long time must have been before his mind--witness his remarks on Dr. Schmerling's work in the caves near Liege; but it had assumed a special significance owing to the famous discovery of flint implements in the valley of the Somme.[133] The whole subject also would have a special interest for Lyell, because he had made Tertiary deposits his special field in stratigraphy, and had worked at this subject downwards, comparing extinct with living forms, so that he had seen more than others of the borderland which blends by an insensible transition the province of the geologist with that of the archeologist. Probably also the thought which he had been giving to the question of the origin of species would bring into no less vivid prominence that of the age and origin of the human race. Be this as it may, he undertook a task comparatively novel, and for the next three years was fully occupied in the preparation of his third great book, "The Antiquity of Man." Travel was necessary for this purpose also; but as the journeys were less lengthy than those already described, and led him for the most part over old ground, it is needless to enter into details. He visited the gravels of the Somme Valley and the caves on the Meuse, besides other parts of Northern France and Belgium,[134] the gravel pits near Bedford, and various localities in England, examining into the evidence for himself, and paying particular attention, not only to the question of man's antiquity, but also to the supposed return of a warmer climate than now prevails after the era of glacial cold. The book was published early in 1863. Naturally its conclusions were startling to many and were vigourously denounced by some; but it was a great success, for it ran through three editions in the course of the year. A fourth and enlarged edition was published in 1873.

The book may seem, from the literary critic's point of view, rather composite in character, and this objection was made in a good-natured form by a writer in the Saturday Review,[135] who called it "a trilogy on the antiquity of man, ice, and Darwin." That, however, is but a slight blemish, if blemish it be, and it was readily pardoned, because of the general interest of the book, the clearness of its style, and the lucidity of its reasoning.

In accordance with his usual plan of work--proceeding tentatively from the known to the unknown--Lyell begins with times nearest to the present era and facts of which the interpretation is least open to dispute. He conducts his reader at the outset to the peat mosses of Denmark, where weapons of iron, bronze, and stone lie in a kind of stratified order; and to those mounds of shells, the refuse heaps of a rude people, which are found on the Baltic shore. Next he places him on the site of the pile-built villages which once fringed the shores of Swiss and Italian lakes. Here weapons of iron, of bronze, and of stone are hidden in peat or scattered on the lake-bed. But these log-built settlements, such as those which Herodotus described at Lake Prasias in Roumelia, are not the only remnants of an almost prehistoric people, for nearer home we find analogous constructions in the crannoges of Ireland--islets partly artificial, built of timber and stone. Lyell then passes on from Europe to the valleys of the Nile and Mississippi, and so to the "carses" of Scotland. In the last case canoes buried in the alluvial deposits, as in the lowland by the Clyde, indicate that some physical changes, slight though they may be, have occurred since the coming of man. But none of these researches lead us back into a very remote past; they keep us still lingering, as it were, on the threshold of history. The weapons which have been described, even if made of stone, exhibit a considerable amount of mechanical skill, for many of them are fashioned and polished with much care, while they are associated with the remains of creatures which are still living at no great distance, if not in the immediate vicinity. Accordingly he conducts his reader, in the next place, to the localities where ruder weapons only have been found, fashioned by chipping, and never polished--namely, to the caves of Belgium and of Britain, of Central and of Southern France, and to the gravel beds in the valleys of the Somme and the Seine, of the Ouse and other rivers of Eastern and Southern England. These furnish abundant evidence that man was contemporary with several extinct animals, such as the mammoth and the woolly rhinoceros, or with others which now inhabit only arctic regions, such as the reindeer and the musksheep, and that the valleys since then have been deepened and altered in contour. This evidence, stratigraphical as well as paleontological, proves that important changes have occurred since man first appeared, not only in climate, but also in physical geography.

The Glacial Epoch is the subject of the second part of the book. Its pages contain an admirable sketch of the deposits assigned to that age in Eastern England, Scandinavia, the Alps, and North America, with special descriptions of the loess of Northern Europe, the drifts of the Danish island of Man, so like those near Cromer, and the parallel roads of Glenroy, which Lyell now supposes to have been formed in a manner similar to that of the little terrace by the Majalen See.

The third part deals with "the origin of species as bearing on man's place in Nature." It is a recantation of the views which he had formerly maintained. In all his earlier writings, including the ninth edition of the "Principles," he had expressed himself dissatisfied with the hypothesis of the transmutation of species, and had accepted, though cautiously and not without allowing for considerable power of variation, that of specific centres of creation. Now, after a full review of the question, he gives his reasons for abandoning his earlier opinions and adopting in the main those advocated by Darwin and Wallace. Nevertheless, through frankly avowing his change of view, he advances cautiously and tentatively, like a man over treacherous ice--so cautiously, indeed, that Darwin is not wholly satisfied with his convert, and chides him good-humouredly for his slow progress and over-much hesitation. But this very hesitation was as real as the conversion: the one was the outcome of Lyell's thoroughly judicial habit of mind, the other was a proof, perhaps the strongest that could be given, of that mind's freshness, vigour, and candour. The book ends with a chapter on "man's place in Nature." On this burning question the author speaks with great caution, but comes to the conclusion that man, so far as his bodily frame is concerned, cannot claim exception from the law which governs the rest of the animal kingdom and he ends[136] with a few words on the theological aspect of the question: "It may be said that, so far from having a materialistic tendency, the supposed introduction into the earth, at successive geological periods, of life--sensation--instinct--the intelligence of the higher mammalia bordering on reason--and, lastly, the improvable reason of man himself, presents us with a picture of the ever-increasing dominion of mind over matter."

FOOTNOTES:

[133] Found by M. Boucher de Perthes, who had published a book on the subject in 1847, and had announced the discovery about seven years earlier; but geologists, for various reasons, were not fully satisfied on the matter till the visit of Messrs. Prestwich and John Evans (now Sir) in 1857.

[134] He went to Florence in 1862, but how far this was for geological work is not stated.

[135] Vol. xv. p. 311.

[136] "Antiquity of Man," chap. xxiv.

CHAPTER XI.

THE EVENING OF LIFE.

The second and third editions of the "Antiquity of Man" were not mere reprints, since new materials were constantly coming in and researches were continued; for during the summer of 1863 Sir Charles was rambling about Wales, visiting the caves of Gower in Pembrokeshire, and of Cefn in Denbighshire, the peats of Anglesea, and the boulder clay and shell-bearing sands near the top of Moel Tryfaen. He also went over to Paris, apparently about this time, to inquire into the authenticity of specimens--bones with notches upon them--which were supposed to prove man contemporaneous with the Cromer Forest Beds of England, and therefore pre-glacial. Shorter journeys were to Osborne (by Royal command), to Suffolk, and to Kent.

While engaged on the above-named book, he had persistently refused more than one position of honour--such as a Trusteeship at the British Museum, to be a candidate for the representation of the University of London in Parliament, even an honorary degree from the University of Edinburgh because he was too busy to undertake the journey. In 1861, also, he seems to have received a warning that he was beginning to grow old, for he became rather seriously unwell, and was ordered to Kissingen in Bavaria to take a course of the waters. But during the same period two acceptable honours were received--namely, the Corresponding Membership of the Institute of France, in 1862, and an order of Scientific Merit from the King of Prussia in the following year.

The years, as must be the case when life's evening shadows are lengthening, begin to be more definitely chequered with losses and with rewards. In his letters, references to the death of friends become frequent. In 1862 Mrs. Horner, Lady Lyell's mother, died, and in 1864 her father, Leonard Horner, with whom, even for some years before becoming his son-in-law, Lyell had been in constant friendly correspondence, passed away in his eightieth year. In the same year Lyell was raised to the rank of baronet, and also occupied the presidential chair at the meeting of the British Association at Bath.

His address deals principally with two topics--one local, thermal springs, especially those of Bath; the other general, the glacial epoch and its relation to the antiquity of man. He refers, however, in the concluding paragraph to the marked change which, within his memory, opinion had undergone, in regard to catastrophic changes and the origin of species, and to the discovery of the supposed fossil Eozoon Canadense in the crystalline Laurentian rocks of Canada. This singular structure appeared to him--as it did to Sir W. Logan, who had brought specimens for exhibition at the meeting--to be a fossil organism,[137] and thus to indicate the existence of living creatures at a much earlier period than hitherto had been supposed. But in stating this opinion he checks himself characteristically with these words: "I will not venture on speculations respecting 'the signs of a beginning,' or 'the prospects of an end' of our terrestrial system--that wide ocean of scientific conjecture on which so many theorists before my time have suffered shipwreck."

The address contains more than one passage that is well worth quotation, but the following has so wide a bearing, and is so significant as to the effects of early influences, that it should not be forgotten:--

"When speculations on the long series of events which occurred in the Glacial and post-Glacial periods are indulged in, the imagination is apt to take alarm at the immensity of the time required to interpret the monuments of these ages, all referable to the era of existing species. In order to abridge the number of centuries which would otherwise be indispensable, a disposition is shown by many to magnify the rate of change in prehistoric times, by investing the causes which have modified the animate and inanimate world with extraordinary and excessive energy. It is related of a great Irish orator of our day, that when he was about to contribute somewhat parsimoniously towards a public charity, he was persuaded by a friend to make a more liberal donation. In doing so, he apologised for his first apparent want of generosity by saying that his early life had been a constant struggle with scanty means, and that 'they who are born to affluence cannot easily imagine how long a time it takes to get the chill of poverty out of one's bones.' In like manner, we of the living generation, when called upon to make grants of thousands of centuries in order to explain the events of what is called the modern period, shrink naturally at first from making what seems to be so lavish an expenditure of past time. Throughout our early education we have been accustomed to such strict economy in all that relates to the chronology of the earth and its inhabitants in remote ages, so fettered have we been by old traditional beliefs, that even when our reason is convinced and we are persuaded that we ought to make more liberal grants of time to the geologist, we feel how hard it is to get the chill of poverty out of our bones."[138]

A presidential address to the British Association is no light task; but, in addition to this, Lyell was now engaged upon a new edition of the "Elements (or Manual) of Geology," which for some time had been urgently demanded; the last edition also of the "Principles"--though 5,000 copies had been printed--was practically exhausted. The former work was cleared off before the end of the year, the book appearing in January, 1865, and the latter was at once taken vigorously in hand, as we see from a letter questioning Sir John Herschel about the earth-pillars on the Rittnerhorn, near Botzen, and on the influence which changes in the shape of the earth's orbit and the position of its axis would have upon climate--a view which had been advocated by Dr. Croll. Lyell, it will be remembered, had originally regarded geographical conditions as the only factors which modified climate, but he was evidently impressed by Croll's argument, and ready, if his mathematics were correct, to admit astronomical changes as an independent, though probably less potent, cause of variation.

The Christmas of 1864 and the following New Year were spent in Berlin, and in the summer of 1865 he had again recourse to Kissingen. Though he writes that the waters "did him neither harm nor good," he was at any rate well enough after the "cure" to undertake a rather lengthy tour with Lady Lyell and his nephew[139] Leonard, in the course of which he examined for himself the wonderful earth-pillars near Botzen, and visited the M錠 jalen See, that pretty lake held up by the ice of the great Aletsch Glacier, in order to see whether it threw any light on the origin of the parallel roads of Glenroy. He was satisfied that it did, for he found there a large terrace "exactly on a level with the col which separates the valley" occupied by the lake from that of the Viesch glacier. On his return to England, he writes a long letter to Sir John Herschel, discussing the origin of these earth-pillars, and making inquiries as to the precise points from which his friend, more than forty years before, had made some elaborate drawings. The expedition, as well as the letter, to quote Lyell's own words, were pretty well for a man who was "battling with sixty-eight years." He complains, however, of little more than occasional attacks of lumbago, and a necessity for taking great care of himself; but his eyes were now more troublesome than they had been, and for the last year he had been driven to avail himself of the services of a secretary,[140] with the result that he seemed to have acquired a new lease of his eyes, and to be able, for ordinary purposes, to use them almost as well as formerly.

After his return from the Continent Sir Charles was working hard at the new edition of the "Principles," which obviously gave him much trouble, for letters still remain which were written to Herschel on questions relating to climate and astronomy; to Hooker, Wallace, and Darwin on the transmutation of species, the distribution and migration of plants and animals, the effects of geographical changes, and even on such matters as the Triassic reptilia of Elgin and Warwickshire, Central India and the Cape. At last the first volume of the new and much-enlarged edition (tenth) was published in November, 1866, the second volume not appearing till 1868. Few men at that time of life could have accomplished such a piece of work, especially if they had been compelled, as Lyell was, to read with the eyes and write with the hands of others. But even now, in regard to field work, he was still able to see things for himself, and, though less vigorous than formerly, to undertake journeys of moderate length. In 1866, in company with his nephew Leonard, he examined the Glacial and late Tertiary deposits of the Suffolk coasts; looked once more at the sections of Jurassic rocks in the Isle of Portland and the neighbourhood of Weymouth, and doubtless speculated on the origin of the Chesil Bank and of the Fleet. One honour fell to him in this year, which, doubtless, only the accident of his long service on the Council had previously kept from him--namely, the Wollaston Medal of the Geological Society.

In 1867 he was strong enough to visit the Paris Exhibition, after which he went to Forfarshire, and attended the meeting of the British Association at Dundee. In the following year he was present at the same gathering in Norwich, besides making various shorter journeys in England and spending September in Pembrokeshire with Lady Lyell and his brother's family,[141] in whose company evidently he took much pleasure.

In the spring of 1868 he was again in the field, examining the splendid plant remains of Eocene age in the neighbourhood of Bournemouth and Poole, and the shallow-water deposits of the Purbeck group ripple-marked and sun-cracked, together with the traces of their ancient forests. Over these he became as enthusiastic as any young geologist. At this time also, apparently, he visited the Blackmore Museum[142] at Salisbury, and himself found reindeer antlers in the neighbouring gravels at Fisherton. In the autumn they again stayed at Tenby with Colonel Lyell's family, when one of the latter was attacked by a serious illness. But Sir Charles was able to take his nephew Leonard to St. David's, and examine the magnificent sections of fossiliferous Cambrian rocks, under the guidance of Dr. H. Hicks, whose name is inseparably connected with the geology of this district.

Comparatively few records are preserved of the last six years of his life; still they are enough to show that his interest in science never flagged. The few letters which have been printed show no signs of declining mental strength. Though his bodily powers had become less vigorous, though his sight was weak, and his limbs were less firm than in the olden times, he was by no means ready to be laid altogether on the shelf. For instance, in the spring of 1869 he went back to the coast of Suffolk and Norfolk, to resume work which he had been unable to complete on his last visit.

Starting at Aldborough, where Pliocene deposits are still exposed, from the Coralline Crag up to the Chillesford group, they examined the coasts by Southwold and Kessingland to Lowestoft, seeing "a continuous section, for miles unbroken, of the deposits from the upper part of the Pliocene to the glacial drift." The Kessingland cliffs afforded good sections of the "Forest Bed," the deposit which on former occasions he had studied in the neighbourhood of Cromer. It was covered by several yards of stratified sand, and that by glacial drift, "with the usual 'boulders' of chalk, flint, lias, sandstone, and other sedimentaries, with crystalline rocks from more distant places." Passing on into Norfolk, they followed this "Forest Bed" and the overlying boulder clay, and they found in the latter, near Happisburgh, some fragments of sea-shells, and one perfect valve of Tellina solidula in a band of gravel, "like a fragment of an old sea-beach," intercalated in the glacial clay. As the origin of this clay has been, of late years, a subject of dispute, it may be interesting to quote Sir Charles's conclusion:--"I suppose, therefore, we must set it down as a marine formation; and underneath it, from Happisburgh to Cromer, comes the famous lignite bed and submarine forest, which must have sunk down to allow of the unquestionable glacial formation being everywhere superimposed."[143]

On revisiting Sherringham (a village about five miles along the coast to the west of Cromer), he found a striking instance of that "sea change" to which in his early days he had called attention. "Leonard and I" (he writes to Sir C. Bunbury) "have just returned from Sherringham, where I found that the splendid old Hythe pinnacle of chalk, in which the flints were vertical, between seventy and eighty feet high, the grandest erratic in the world, of which I gave a figure in the first edition of my "Principles," has totally disappeared. The sea has advanced on the lofty cliff so much in the last ten years, that it may well have carried away the whole pinnacle in the thirty years which have elapsed since our first visit."

Another letter, bearing date in the next month, to Darwin shows that in his seventy-second year his mind was fresh and keen as ever. It discusses an article written by Wallace in the Quarterly Review, and indicates the difference in regard to natural selection between Lyell's own standpoint and that of his correspondent. The following extract may serve to show the general tenor of the remarks:--"As I feel that progressive development in evolution cannot be entirely explained by natural selection, I rather hail Wallace's suggestion that there may be a Supreme Will and Power, which may not abdicate its functions of interference, but may guide the forces and laws of Nature." In another passage he refers, to a controversy which had been recently started by Professor (afterwards Sir A.) Ramsay, and over which geologists have been fighting ever since--viz. whether lake-basins are excavated by glaciers. The passage is worth quoting, for it puts the issue in a form which after a quarter of a century is virtually unchanged:--

"As to the scooping out of lake-basins by glaciers, I have had a long, amicable, but controversial correspondence with Wallace on that subject, and I cannot get over (as, indeed, I have admitted in print) an intimate connection between the number of lakes of modern date and the glaciation of the regions containing them. But as we do not know how ice can scoop out Lago Maggiore to a depth of 2,600 feet, of which all but 600 is below the level of the sea, getting rid of the rock supposed to be worn away as if it was salt that had melted, I feel that it is a dangerous causation to admit in explanation of every cavity which we have to account for, including Lake Superior. They who use it seem to have it always at hand, like the 'diluvial wave or the wave of translation,' or the 'convulsion of nature or catastrophe' of the old paroxysmists."[144]

In the summer he took a longer tour, going first to Westmoreland and then to Forfarshire; after which, in company with Lady Lyell and his nephew, he went to see the old rocks of Ross-shire, above Inchnadamff and Ullapool, and, as he returned, once more visited the parallel roads of Glenroy.

But, in the meantime, notwithstanding the difficulties mentioned above, he still kept working at his books. He was now engaged in modifying the "Elements of Geology." Of this, to quote the preface afterwards published, he had published "six editions between the years 1838 and 1865, beginning with a small duodecimo volume, which increased with each successive edition, as new facts accumulated, until in 1865 it had become a large and somewhat expensive work." He therefore determined, in accordance with the advice of friends, "to bring the book back again to a size more nearly approaching the original, so that it might be within the reach of the ordinary student." This was done by the omission of certain theoretical discussions and all such references to Continental geology as were not absolutely necessary.[145]

In 1870 Sir Charles continued to travel, though within the limits of these islands, for he made one journey along the coast of North Devon, and a second one to Scotland, in the course of which he visited the Isle of Arran, and on his return halted first at Ambleside and then at Liverpool, to attend the meeting of the British Association, which began on the 14th of September. The following year he paid an April visit to Tintagel, the Land's End, and other parts of Cornwall, and in the summer went to the North of England. Writing from Penrith to Sir C. Bunbury, he remarks "that he had much enjoyed his 'tour of inspection,' and had tried to make it a tour of rest, which is difficult." Naturally so, for he had been working his way from Buxton on the look-out for glacial deposits and studying especially the stratified drifts on the hills east of Macclesfield, 1,200 feet above the sea. His remarks on these show that he appreciated fully both the significance of the marine fossils which they contain and the theoretical difficulties caused by the absence of such remains in other deposits, whether in Derbyshire or the Lake District, or in the lowland between this locality and Moel Tryfaen, seventy-four miles away.

The tenth edition of the "Principles" had been quickly sold, and Sir Charles was now employed in the preparation of another one. In this less change was necessary than on the last occasion; still, the rapid increase of knowledge, more especially in regard to the temperature and currents of the sea, obliged him to make considerable alterations in the parts which dealt with these subjects and with questions of climate, so that he recast or rewrote five chapters.

It was published in January, 1872; and in the summer of that year, no doubt in view of a new edition of the "Antiquity of Man," he went to the south of France, with Lady Lyell and Professor T. M'K. Hughes, to examine the Aurignac cave. Here several human skeletons had been discovered some years before, apparently entombed with the bones of various extinct mammals, such as the cave-bear and lion, the mammoth and woolly rhinoceros--in short, with a fauna characteristic of the paleolithic age. But was this really the date of the interment? Some distinguished geologists were of opinion that, though the cave had been then occupied by wild beasts, its floor had been disturbed, and the corpses buried in neolithic times. On this point Lyell was unable to obtain conclusive evidence, and was obliged to confine himself to a statement of the facts and arguments on either side of the question.[146]

Shortly after the publication of this new edition of the "Antiquity of Man" in January, 1873, an unexpected and irreparable bereavement darkened the evening of his days. On April 24th Lady Lyell, the companion and helpmate of forty years, was taken from him after a few days' illness from an inflammatory cold.[147] The shock was the more severe because the loss was so unforeseen. Lady Lyell was twelve years his junior, and had always enjoyed good health[148]--"youthful and vigorous for her age," as he writes--so that he "never contemplated surviving her, and could hardly believe it when the calamity happened." He bore the blow bravely, consoling himself by reflecting that the separation, at his age--nearly seventy-six--could not be for very long, and, as he writes to Professor Heer, of Zurich, endeavouring, "by daily work at my favourite science, to forget as far as possible the dreadful change which this has made in my existence."

Lady Lyell was a woman of rare excellence. "Strength and sweetness were hers, both in no common degree. The daughter of Leonard Horner, and the niece of Francis Horner, her own excellent understanding had been carefully trained, and she had that general knowledge and those intellectual tastes which we expect to find in an educated Englishwoman; and from her childhood she had breathed the refining air of taste, knowledge, and goodness. Her marriage ... gave a scientific turn to her thoughts and studies, and she became to her husband, not merely the truest of friends and the most affectionate and sympathetic of companions, but a very efficient helper. She was frank, generous, and true; her moral instincts were high and pure; she was faithful and firm in friendship; she was fearless in the expression of opinion without being aggressive; and she had that force of character and quiet energy of temperament that gave her the power to do all that she had resolved to do.... She had more than a common share of personal beauty; but had she not been beautiful she would have been lovely, such was the charm of her manners, which were the natural expression of warmth and tenderness of heart, of quick sympathies, and of a tact as delicate as a blind man's touch."[149]

He was not, however, left to bear in solitude the burden of darkening sight and of a desolated home. His eldest sister, Miss Lyell, came from Kinnordy to take care of his house and watch over him in these last years with an affectionate devotion; and in her company and that of Professor Hughes he even carried out the plan, which had been already in contemplation, of once more going on to the Continent and of visiting Professor Heer, at Zurich.

He worked on, as well as slowly increasing infirmities allowed, after his return to England, fully occupied in preparing a second edition of the "Student's Elements" and a new one of the "Principles."[150] In June, 1874, he again visited Cambridge, this time to receive the degree of LL.D.--an honour which that University had been strangely slow in conferring upon him.[151] It was then too evident that his strength was declining, for he became quickly fatigued by any exertion of body or mind; nevertheless, he was able soon afterwards to make once more the journey to Forfarshire, and to visit there several of his earlier geological haunts. In some of these little excursions he had as his companion Mr. J. W. Judd,[152] with whose recent researches into the ruined volcanoes of Tertiary age and the yet earlier stratified rocks in the Western Isles of Scotland Sir Charles was hardly less interested than he would have been in the days when the "Principles" was a new book. Three or four letters written about this time have been printed[153] which show, from their vigour and freshness, that the mind was still keen and bright, though the bodily machinery was becoming outworn.

After his return to town he even ventured, on November 5th, to dine at the Geological Club,[154] of which he had been a member from its foundation, on its fiftieth anniversary meeting, and "spoke with a vigour which surprised his friends."

The tale, however, is nearly told; the sands of life were running low. "His failing eyesight and other infirmities now began to increase rapidly, and towards the close of the year he became very feeble. But his spirit was ever alive to his old beloved science, and his affectionate interest and thought for those about him never failed. He dined downstairs on Christmas Day with his brother's family, but shortly after that kept to his room."

On February 22nd, 1875, Charles Lyell entered into his rest. The end may have been slightly accelerated by two causes--one, the death, from inflammation of the lungs, after a short illness, of his brother,[155] Colonel Lyell, who, up to that time, had visited him almost daily; the other, the shock given to his enfeebled system by accidentally falling on the stairs a few weeks before. But in no case could it have been long delayed; the bodily frame was outworn; the hour of rest had come.

His fellow-workers in science felt unanimously that but one place of sepulture was worthy to receive the body of Charles Lyell--the Abbey of Westminster, our national Valhalla. A memorial, bearing many important signatures, was at once presented to Dean Stanley, who gave a willing consent, and the interment took place with all due solemnity on Saturday the 27th. The grave was dug in the north aisle of the nave, near that of Woodward, one of the pioneers of British geology and the founder of the chair of that science in the University of Cambridge. It is marked[156] by a slab of Derbyshire marble, which bears this inscription:--

CHARLES LYELL, BARONET, F.R.S., AUTHOR OF "THE PRINCIPLES OF GEOLOGY." BORN AT KINNORDY, IN FORFARSHIRE, NOVEMBER 14, 1797; DIED IN LONDON, FEBRUARY 22, 1875.

THROUGHOUT A LONG AND LABORIOUS LIFE HE SOUGHT THE MEANS OF DECIPHERING THE FRAGMENTARY RECORDS OF THE EARTH'S HISTORY IN THE PATIENT INVESTIGATION OF THE PRESENT ORDER OF NATURE, ENLARGING THE BOUNDARIES OF KNOWLEDGE AND LEAVING ON SCIENTIFIC THOUGHT

AN ENDURING INFLUENCE.

"O LORD, HOW GREAT ARE THY WORKS, AND THY THOUGHTS ARE VERY DEEP. PSALM XCII. 5.

Sir Charles, by his will, left to the Geological Society of London the die, executed by Mr. Leonard Wyon, of a medal to be cast in bronze, and awarded annually to some geologist of distinction, whether British or foreign. He further left a sum of two thousand pounds, free of legacy duty, to the Society, in trust, the interest of it to be applied as follows:--Not less than one-third of it to accompany the medal, and the remainder to be given, in one or more portions, for the furtherance of the science. Sir Charles was succeeded in the family estates by his nephew Leonard, the eldest son of Colonel Lyell, who lives at Kinnordy, but has rebuilt the house. He was created a baronet in 1894.

FOOTNOTES:

[137] The nature of Eozoon, whether it be the remains of a foraminifer of unusual size and peculiar habit of growth, or merely a very exceptional arrangement of its constituent minerals, has been since the above-named date a fruitful subject of controversy. For some years the balance of opinion was in favour of an organic origin; now it seems to be distinctly tending in the other direction.

[138] Report of Brit. Assoc., 1864, p. xxiv.

[139] Colonel Lyell's eldest son, the present baronet.

[140] He was fortunate in obtaining the help of Miss Arabella Buckley, a lady of congenial tastes in literature and science.

[141] The relationship was unusually close, for Colonel Lyell had married another Miss Horner.

[142] For a description of this fine collection of prehistoric antiquities, see "Flint Chips," by E. T. Stevens, 1870.

[143] Life, Letters, and Journals, ii. p. 440.

[144] Life, Letters, and Journals, ii. p. 443.

[145] The book, thus abbreviated, and entitled "The Student's Elements of Geology," was published in 1871. A second edition appeared in February, 1874; a third, revised by Mr. Leonard Lyell and others, in 1878; and a fourth, edited by Prof. P. M. Duncan, in 1885.

[146] "Antiquity of Man" (fourth edition), chap. vii.

[147] She had been suffering from influenza, but had accompanied her husband and nephews to Ludlow at the beginning of the month. They became uneasy at her increasing debility, and returned to town on the 14th ("Life, Letters, and Journal of Sir C. Bunbury," iii. p. 9).

[148] He mentions, on January 5th, 1856, that she had not been well enough to breakfast with him, "for the second time only since our marriage."

[149] Quoted from an obituary notice by G. S. Hillard, Esq., in the Boston (U.S.) Daily Advertiser (printed in Life, Letters, and Journals, ii. p. 467).

[150] This was published after his death. He had completed one volume; the other was revised by his nephew Leonard.

[151] About the same time he was admitted to the freedom of the Turners' Company in the City of London.

[152] Now Professor Judd, F.R.S., of the Royal College of Science, South Kensington.

[153] Life, Letters, and Journals, ii. pp. 453-459.

[154] The Club consists of a certain number of Fellows of the Geological Society, who dine together before the evening meetings.

[155] His brother Thomas, who had retired from the Navy with the rank of captain, had died (unmarried) some years before at the jointure house (Shiel Hill), Kinnordy, where he had resided with one of his sisters.

[156] A marble bust, a copy by Theed of the original executed by Gibson, is placed near the grave.

CHAPTER XII.

SUMMARY.

In stature, Sir Charles Lyell[157] was rather above the middle height, somewhat squarely built, though not at all stout, with clear-cut, intellectual features, and a forehead, broad, high, and massive. He would have been a man of commanding presence, if his extremely short sight had not obliged him to stoop and peer into anything he wished to observe. This defect, in addition to the weakness of his eyes was a serious impediment in field work. As Professor Ramsay remarked in 1851, after spending a few days with him in the south of England, he required people to point things out to him, and would have been unable to make a geological map, "but understood all when explained, and speculated thereon well."[158] This defect of sight, according to Sir J. W. Dawson, who had been his companion in more than one excursion in Canada, was at times even a source of danger. The expression of his face was one of thoughtful power and gracious benignity.[159] "In his work, Lyell was very methodical, beginning and ending at fixed hours. Accustomed to make use of the help of others on account of his weak sight, he was singularly unconscious of outward bodily movement, though highly sensitive to pain. When dictating, he was often restless, moving from his chair to his sofa, pacing the room, or sometimes flinging himself full length on two chairs, tracing patterns on the floor, as some thoughtful or eloquent passage flowed from his lips. But though a rapid writer and dictator, he was sensitively conscientious in the correction of his manuscript, partly from a strong sense of the duty of accuracy, partly from a desire to save his publisher the expense of proof corrections. Hence passages once finished were rarely altered, even after many years, unless new facts arose."

The characteristic with which anyone who spent some time in Charles Lyell's company was most impressed, was his thirst for knowledge, combined with a singular openness, and perfect fairness of mind. He was absolutely free from all petty pride, and from "that common failing of men of science, which causes them to cling with such tenacity to opinions once formed, even in the face of the strongest evidence."[160] Ramsay wrote of him,[161] "We all like Lyell much; he is anxious for instruction, and so far from affecting the bigwig, he is not afraid to learn anything from anyone.[162] The notes he takes are amazing." No man could have given a stronger proof of candour and plasticity of mind and of his care for truth alone than Lyell did in dealing with the question of the origin of species. From the first he approached it without prejudice. So long as the facts adduced by Lamarck and others appeared to him insufficient to support their hypotheses, he gave the preference to some modification of the ordinarily accepted view--that a species began in a creative act--but after reading Darwin's classic work,[163] and discussing the subject in private, not only with its author, but also with Sir J. Hooker and Professor Huxley, he was convinced that Darwin was right in his main contention, though he held back in regard to certain minor points, for which he thought the evidence as yet insufficient. Of his conduct in this matter, Darwin justly wrote: "Considering his age, his former views, and position in society, I think his action has been heroic."[164] Dean Stanley, in the pulpit of Westminster Abbey, on the Sunday following the funeral, summed up in a few eloquent sentences the great moral lesson of Lyell's life. "From early youth to extreme old age it was to him a solemn religious duty to be incessantly learning, fearlessly correcting his own mistakes, always ready to receive and reproduce from others that which he had not in himself. Science and religion for him not only were not divorced, but were one and indivisible."[165]

To ascertain the truth, and to be led by reason not by impulse, that was Lyell's great aim. Sedgwick once[166] criticised his work in terms which, in one respect, seem to me curiously mistaken: "Lyell ... is an excellent and thoughtful writer, but not, I think, a great field observer ... his mind is essentially deductive not inductive." The former criticism, as has been already admitted, is just, but the latter, pace tanti viri, seems to me the reverse of the truth. Surely there never was a geologist whose habits and methods were more strictly inductive than Lyell's. He would spare no pains, and hardly any expense, to ascertain for himself what the facts were; he abstained from drawing any conclusion until he had accumulated a good store; he compared and marshalled them, and finally adopted the interpretation with which they seemed most accordant. This interpretation, however, would be modified, or even rejected, if new and important facts were discovered. Surely this is the method of induction; surely this is the mode of reasoning adopted by Darwin and by Newton, and even by Bacon himself. But Sedgwick, great man as he was, almost unrivalled in the field, more brilliant, though less persevering than Lyell, was not always quite free from prejudices; and it may be noted that he more than once stigmatises an opinion which he dislikes by declaring it not to be in accordance with inductive methods. Sir Joseph Hooker's judgment was far more accurate: "One of the most philosophical of geologists, and one of the best of men"[167]; or that of Charles Darwin himself: "The science of geology is enormously indebted to Lyell--more so, as I believe, than to any other man who ever lived."[168]

Lyell felt a keen interest in the broader aspect of political questions, and this not only in his own country,[169] though he took little or no share in party struggles, for the vulgarity of the demagogue and the coarseness of the hustings were offensive to a man of such refinement. His opinions harmonised with his scientific habits of thought, always progressive, but never extravagant. He was in favour of greater freedom in education, of the restriction of class privileges, and of an extension of the franchise, but he saw clearly that anything like universal suffrage, as the world is at present constituted, would only mean giving a preponderating influence to those least competent to wield it; that is, to the more ignorant and easily deluded. As in such cases the glib tongue would become more potent than the voice of reason, the demagogue than the statesman, he feared that the standard of national honour would be almost inevitably lowered, and national disaster be a probable result. That all men are equal and entitled to an equal share in the government--a dogma now regarded in some circles as almost sacred--would have been repudiated by him with the quiet scorn of a man who prefers facts to fancies, and inductive reasoning to sentimental rhapsody. A partisan he could not be, for he saw too clearly that in political matters truth and right were seldom a monopoly of any side, and though by no means wanting in a certain quiet and restrained enthusiasm, he had almost an abhorrence of fanaticism. One example may serve for many, to indicate the way in which he regarded both this spirit and any difficult question. Naturally he had a strong dislike to slavery; he fully recognised the injustice and wrong to the negro, and the evil effects upon the master. Nevertheless, after visiting the Southern States, and giving the impressions of his journey, he thus expresses himself: "The more I reflected on the condition of the slaves, and endeavoured to think on a practical plan for hastening the period of their liberation, the more difficult the subject appeared to me, and the more I felt astonished at the confidence displayed by so many anti-slavery speakers and writers on both sides of the Atlantic. The course pursued by these agitators shows that, next to the positively wicked, the class who are usually called 'well-meaning persons' are the most mischievous in society." He then points out how a strong feeling against slavery had been springing up in Virginia, Kentucky, and Maryland; how the emancipation party had been gaining ground, and slavery steadily retreating southwards, but "from the moment that the abolition movement began, and that missionaries were sent to the Southern States, a reaction was perceived--the planters took the alarm--laws were passed against education--the condition of the slave was worse, and not a few of the planters, by dint of defending their institutions against the arguments and misrepresentations of their assailants, came actually to delude themselves into a belief that slavery was legitimate, wise, and expedient--a positive good in itself."[170] At a subsequent period he speaks of Mrs. Beecher Stowe's famous book, "Uncle Tom's Cabin," as "a gross caricature." But in the great struggle between the Northern and Southern States, his sympathies went with the former. It was the fairness of his criticisms, and his hearty appreciation of the good side in American institutions, that won him many friends and made his books welcome on that side of the Atlantic.

Lyell's views on religious questions accorded, as might be expected, with the general bent of his mind. He was a member of the Church of England,[171] appreciated its services, the charm of music, and the beauty of architecture, but he failed to understand why nonconformity should entail penalties, whether legal or social. His mind was essentially undogmatic; feeling that certainty was impossible in questions where the ordinary means of verification could not be employed, he abstained from speculation and shrank from formulating his ideas, even when he was convinced of their general truth.

He was content, however, to believe where he could not prove, and to trust, not faintly, the larger hope. So he worked on in calm confidence that the honest searcher after truth would never go far astray, and that the God of Nature and of Revelation was one. He sought in this life to follow the way of righteousness, justice, and goodness, and he died in the hope of immortality.

As he disapproved of any approach to persecution on the ground of religion, so he objected strongly to the exclusive privileges which in his day were enjoyed by the Church of England, especially to its virtual monopoly of education. On this point he several times expresses himself in forcible terms; as, for instance, in these words: "The Church of England ascendency is really the power which is oppressive here, and not the monarchy, nor the aristocracy. Perhaps I feel it too sensitively as a scientific man, since our Puseyites have excluded physical science from Oxford. They are wise in their generation. The abject deference to authority advocated conscientiously by them can never survive a sound philosophical education."[172] To this party-- or to the "Catholic movement," as it is now often called--in the Church of England, Lyell had a strong dislike; he deemed their claims to authority unwarrantable, their practices in many respects either childish or superstitious.

As we have endeavoured to bring out in the course of this volume the guiding principles of Lyell's work, a brief recapitulation only is needed as a conclusion. That work was regulated by two maxims: the one, "Go and see"; the other, "Prefer reason to authority." To the first maxim he gave expression more than once, while he was always inculcating it by example. Imitating the well-known saying of Demosthenes in regard to oratory, he emphatically declares that in order to form comprehensive views of the globe, the first, the second, and the third requisite is "travel."[173] What he preached, he practised; about a quarter of the last fifty years of his life must have so been spent. Of the second maxim also he was a living example. It was his practice not only to see for himself, but also to judge for himself, in all questions other than those necessarily reserved for specialists; his rule, that thought should be free from the fear of man, but subject to the laws of reasoning. As a young man he had advocated, almost single-handed, scientific views which were unpopular alike with the older authorities in geology and with the supposed friends of religion; he had protested against the invocation of catastrophic destruction and cataclysmal flood in order to clear away difficulties in the past history of the earth; in other words, against an appeal to miracle, when a cause could be found in the existing order of Nature; and he had disputed the right of any priesthood, whether Romanist or Protestant, to hold the keys of knowledge. He vindicated, against all corners, his claim--nay, his birthright--to sit, as an earnest student, at the feet of Nature to listen and to learn, as she chose to teach, whether by the acted drama of the living world or by the silent record of the rocks. He was, in short, more observer than theorist, more philosopher than poet, more a servant of reason than a dreamer of dreams.

His example is one well worthy of remembrance at the present epoch. The "whirligig of time" has brought its revenges, and has introduced into geology a class of students almost unknown in the days when Lyell was in his vigour. The developments of mineralogy and paleontology, helpful and valuable as they have been by making geology more of an exact science and, in some cases, substituting order for confusion, have tended to produce students very familiar with the apparatus of a laboratory or the collections of a museum, but not with the face of the earth. This, in itself, would not be necessarily hurtful, because the field of geology is so wide that there is room for all; but it leads sometimes to an undue exaltation of trifles, to an over-estimation of the "mint, anise, and cummin" of science, to a waste of time upon what is called the literature of the subject. This last often means either searching much chaff for a few grains of wheat, or spending much labour with the hope of discovering whether A or B was the first to confer a name upon a species; the priority perhaps being only of a few months, and that name neither particularly appropriate nor euphonious. Partly from this, partly from other causes, the importance, nay, the absolute necessity of travel, for the education of a geologist is now too often forgotten. In this science there are many questions--some of them almost fundamental--for which no perquisitions in a library, no research in a laboratory, no studies in a museum, however conscientiously patient and painstaking they may be, can be accepted as an adequate preparation; questions in which Nature is at once the best book, the best laboratory, and the best museum, and experience is the only safe teacher. What would Lyell have said to men--and such might now be named--who undertook to discuss wide geological problems with the most limited experience who, for example, posed as authorities upon what ice can or cannot do, without having even seen a glacier or speculated on the most intricate questions in petrology without having studied more than some corner of this island, or, indeed, without any precise knowledge of that? Would not he--averse as he was to speaking severely--have censured them for talking about things which they could not possibly understand, and for darkening counsel by words without knowledge?

Lyell, no doubt, had exceptionally favourable opportunities. The eldest son of a wealthy man--who contentedly acquiesced in his seeking fame rather than fortune, and supplied him with the necessary funds--his time was his own, as he had not only enough for his ordinary wants, but also could afford to travel as much as he desired. His social position was sufficiently good to facilitate his access to those who had already attained to eminence. He was blessed with a sympathetic and helpful wife, and they had no children. Thus they were perfectly free, both in the disposal of their time at home and in their peregrinations abroad. Besides these things they both enjoyed good health. Lyell's constitution was not, indeed, so robust that he could take liberties; he had to be careful about "cakes and ale," and to lead a fairly regular life,[174] but by so doing he was able to be always in good condition for his work. His eyes, in fact, were his only trouble and who is there who has not got his own "thorn in the flesh"? Lyell also was happy in all his domestic relations. His letters indicate that all the family--on both sides--were on affectionate terms, and contain few references to anxieties and troubles, such as the sickness and death of those dear to him, until his life approached the period when such trials become inevitable.

Thus free from the impediments which have beset many other men of marked ability, such as weak health and physical suffering, the wearing anxiety of an invalid wife or a sickly family, the harassing cares of pecuniary losses or of an insufficient income, Lyell had an exceptional chance. But other men have the same and do not use it; they are crippled by this burden or diverted by that allurement, and "might have been" too often becomes their epitaph. Lyell never faltered in the course which, comparatively early in life, he had marked out for himself. With that steady persistency and quiet energy which are characteristic of the Lowland Scot, he put aside all temptations and everything which threatened to interfere with his work. While neither recluse nor hermit, neither churlish nor unsociable, nay, while thoroughly enjoying witty and intellectual society, he allowed nothing to distract him from his main purpose. Convinced that there was a work which he could do, and a name which he could win, he was willing, for sake of this, to run risks and to make sacrifices. He did not indeed despise fame, but he never condescended to unworthy arts to obtain it; he held that the labourer was worthy of his hire, but with him it was always "the work first, and the wage second," whether that were coined gold or laurel wreath. He was singularly free from all petty jealousies, and ready to learn from all who could teach him anything, but he was no weakling, swayed by every breath of wind, for he reached his conclusions slowly and cautiously, and never stopped to ask whether they would be popular. "Forward, for truth's sake," that was the motto of his life.

In yet another way was Lyell felix opportunitate vit? In his days, geology might be compared to a country which had been for some time discovered but was not yet explored. Settlements had been established here and there; in their neighbourhood some ground had been cleared, and a firm base of operations had been secured, but around and beyond was the virgin forest, the untrodden land. At almost every step the traveller met with some fresh accession to his knowledge or a new problem to solve. He could feel the allurement of expectation or the joy of discovery even in countries otherwise well known; where now he can hope only to pick up some tiny detail or to plunge into some interminable controversy. If he now desires "fresh fields and pastures new," he must wander beyond the limits of civilised lands; for within these every crag is hammer-marked, and the official geologist is at work making maps. But not only this, Lyell lived in the days when the literature of his science was of very modest dimensions. This had its obvious drawbacks, but it had also its advantages, which, perhaps, were more than compensations. At the present day the conscientious student is in danger of being overwhelmed by the mass of papers, pamphlets and books, from all lands and in all languages--which he is expected, if not to read, at least to scramble through before venturing to write on any subject. Fifty years ago it required a very limited amount of study--often only a few hours' research--to put the geologist in possession of all that was known, so that he approached his theme very much as a mathematician attacks a problem. This burden of scientific literature, seeing that life is short and human strength is limited, threatens to stifle the progress of science itself, and we can hardly venture to expect that any more great generalisations will be made in geology or paleontology, unless a man arise who is daring enough to subordinate reading to thinking, and so strong in his grasp of principles that he can make light of details.

It has been sometimes said that Lyell was not an original thinker. Possibly not; vixere fortes ante Agamemnona is true in science no less than in national history; there were mathematicians before Newton, philosophic naturalists before Darwin, geologists before Lyell. He did not claim to have discovered the principle of uniformity. He tells us himself what had been done by his predecessors in Italy and in Scotland: but he scattered the mists of error and illusion, he placed the idea upon a firm and logical basis; in a word, he found uniformitarianism an hypothesis, and he left it a theory. That surely is a more solid gift to science, a better claim to greatness, than any number of brilliant guesses and fancies, which, after coruscating for a brief season to the amazement of a gaping crowd, explode into darkness, and are no more seen. But to a certain extent Lyell has thrown his own work into the shade. The fame of his books causes his numerous scientific papers[175] to be overlooked; particularly his contributions to the history of coalfields and to the classification of the Tertiary deposits. Moreover, into these books he was constantly incorporating new and original matter. We may be fairly familiar with the "Principles" and the "Elements," but we fail to realise until we have read his "Life" and the accounts of his two tours in America how much those books are made up from the results of actual experience and personal study in the field.

It has been also said that Lyell carried the principle of "uniformity" a little too far. But, suppose we concede this, does it amount to more than the admission that he was human? It is almost inevitable that the discoverer or prophet of a great truth, who has to encounter the storm and stress of controversy, should state his case a little too strongly, or should overlook some minor limitation. Suppose we grant that Lyell was a little too lavish in his estimate of the time at the disposal of geologists. The physicist had not then intervened, with arguments drawn from his own science, to insist that neither earth nor sun can reckon their years by myriads of myriads, and even now this controversy cannot be regarded as closed. Suppose we grant that in accepting Hutton's dictum, "I find in the earth no signs of a beginning," Lyell was misled by appearances,[176] which have since proved to be delusive, and that facts, so far as they go, point rather in the contrary direction. Well, this point also is not yet to be regarded as settled; and of one thing, at any rate, we may be sure, that if Lyell were now living he would frankly recognise new facts, as soon as they were established, and would not shrink from any modification of his theory which these might demand. Great as were his services to geology, this, perhaps, is even greater--for the lesson applies to all sciences and to all seekers after knowledge--that his career, from first to last, was the manifestation of a judicial mind, of a noble spirit, raised far above all party passions and petty considerations, of an intellect great in itself, but greater still in its grand humility; that he was a man to whom truth was as the 'pearl of price,' worthy of the devotion and, if need be, the sacrifice of a life.

FOOTNOTES:

[157] In this paragraph I have ventured to quote largely, and more or less verbatim, from the words of Miss Buckley (Lyell's secretary) in the article on his life, written by my friend Professor G. A. J. Cole, in the "Dictionary of National Biography," vol. xxxiv.

[158] "Life of Sir A. Ramsay," by Sir A. Geikie, chap. v.

[159] Vidi tantum, when his powers were beginning to fail, but it is this expression which is stamped on my mind as characteristic of the face in Charles Lyell, and, I may add, also in Charles Darwin.

[160] J. W. Dawson, cited in the " Dictionary of National Biography."

[161] Ut supr?

[162] I may add my own testimony. When the second edition of the "Student's Elements" was passing through the press. I ventured to write to him about one or two petrological details, which I thought might be more precise. Though at that time I had published but few papers, I received more than one kind letter with the request that I would read some of the proof-sheets of the book and suggest alterations.

[163] "The Origin of Species," published in 1859.

[164] "Life and Letters of C. Darwin," ii. p. 326.

[165] Quoted in Life, Letters, and Journals, ii. p. 461.

[166] In 1865. "Life and Letters of Sedgwick," ii. p. 412.

[167] "Life, Letters, and Journal of Sir C. Bunbury," iii. p. 66.

[168] "Life and Letters of C. Darwin," i. p. 76.

[169] He maintained for many years an interesting correspondence with Mr. G. Ticknor, of Boston, U.S.A., in which he often discusses political questions, both British and American.

[170] "Travels in North America," chap. ix.

[171] In the later part of his life he appears to have sympathised more with the "Unitarians," for he attended the services at Dr. Martineau's chapel in Little Portland Street, though I am not aware that he formally seceded from the Church of England.

[172] Life, Letters, and Journals, vol. ii. pp. 82-127. It must however, be remembered that the High Church party were not alone in their opposition; indeed, after a time, they were more tolerant of geologists than the extreme "Evangelical" school. I have some cuttings from the Record newspaper, dated about 1876, which are interesting examples of narrow-minded ignorance and theological arrogance.

[173] Life, Letters, and Journal, i. p. 233. "Principles," i. 69 (eleventh edition).

[174] He admits that when Lord Enniskillen and Murchison had seduced him, after a Geological Society meeting, to partake of pterodactyl (woodcock) pie and drink punch into the small hours, his work suffered for four or five days afterwards.

[175] These were about seventy-six in number, the great majority written prior to the last twenty years of his life.

[176] Such as the seeming intercalation of crystalline schists with fossiliferous rocks, or the immediate sequence of the two.

Printed in Great Britain
by Amazon